REPORTS OF DISTRICT MINING ENG......

In 1919–1920 the Mining Bureau was organized into four main geographic divisions, with the field work delegated to a mining engineer in each district, working out from field offices that were established in Redding, Auburn, San Francisco and Los Angeles, respectively. This move brought the office into closer personal contact with operators, and it has many advantages over former methods of conducting field work, including lower traveling-expense bills for the Bureau's engineers. In 1923 the Redding and Auburn field offices were consolidated and moved to Sacramento.

The Redding office was reestablished in 1928, and the boundaries of each district adjusted. The counties now included in each of the four divisions and the location of the branch offices are shown on the accompanying outline map of the state. (Frontispiece.)

Reports of mining activities and development in each district, prepared by the District Engineer, will continue to appear under the proper field division heading.

REDDING FIELD DISTRICT

CHAS. VOLNEY AVERILL, Mining Engineer

MINES AND MINERAL RESOURCES OF SISKIYOU COUNTY

Geography

Siskiyou County, one of the northernmost in the state, borders on the state of Oregon for a distance of 116 miles and is from 60 to 70 miles wide. It is bounded on the west by the coast counties of Humboldt and Del Norte, on the south by Trinity and Shasta, and on the east by Modoc. The region of interest to the miner is the portion lying west of the lava sheet, the western boundary of which coincides roughly with the Oregon branch of the Southern Pacific Railroad, traversing the county from north to south a little east of its center. The total area of the county is 6256 square miles, and the population is 25,500 persons by the census of 1930. This is concentrated largely along the railroad, at the county seat, and in Scott Valley.

The western mineral-bearing section is a succession of high mountains and deep canyons, forming the drainage basin of Klamath River and its tributaries. The Klamath follows a crooked course across the northern part of the county, receiving Shasta and Scott rivers, the principal streams draining the central part of the county; then turning south near Happy Camp where Indian Creek enters, it flows through a deep canyon, receiving the waters of Salmon River at Somes Bar on the southwest county line. The Salmon drains the rugged mineral-bearing area of the southwestern part of the county. Level land is in Shasta Valley and Scott Valley, both lying at an elevation of about 3000 ft. above sea level; also a few gravel bars along Klamath River between Hornbrook and Happy Camp, and such small valleys as those at the mouths of Seiad and Indian creeks, which accommodate the settlements of Seiad and Happy Camp.

Climate is indicated by the following table.[1] Orleans and Weitchpec are a little to the west in Humboldt County.

Place	Elevation, feet	Average annual precipitation, inches	Average seasonal fall of snow, inches	Mean temperature F°	Highest temperature F°	Lowest temperature F°
Dunsmuir	2285	57.06	70.5	---	---	----
Gilta	3300	53.24	108.5	---	---	----
Happy Camp	1132	39.91	----	---	---	----
Hornbrook	2154	14.47	----	---	105	—12
Macdoel	4258	15.07	44.3	43.7	102	—15
McCloud	3270	45.18	119.7	48.4	101	— 3
Montague	2450	12.21	----	51.2	108	—10
Mt. Shasta City	3555	35.59	117.3	48.8	108	— 9
Orleans	520	48.66	----	59.4	117	18
Walla Walla Creek	2570	24.80	----	---	---	----
Weitchpec	1700	73.96	51.5	52.4	102	17
Yreka	2625	17.49	----	50.8	108	— 4

While figures on snowfall are lacking for some of the points mentioned, that is not to be taken as an indication that no snow falls, as there is some snowfall in all parts of the county; and at the higher elevations the depth of fall is greater than any of the figures given in the table. The figures appearing under the name, Walla Walla Creek, represent records taken at a station of that name near Fort Jones, also at Fort Jones and at Scott Valley, in the years from 1853 to 1892.

GEOLOGY AND MINERAL RESOURCES

A map of the areal geology and a description of the general and economic geology of a part of Siskiyou County appeared in State Mineralogist's Report XXVII,[2] chapter for January, 1931, copies of which are available at offices of this division at 25¢ per copy plus 10¢ postage. The following descriptions of mining districts are reprinted in part from State Mineralogist's Report XXI,[3] chapter for October, 1925, which is now out of print.

The county has been celebrated in the past particularly because of its placer mines. Geologically these placer deposits differ from those in the east-central part of the state. In northwestern California the ancient auriferous gravels most actively mined are found in the form of a succession of terrace or bench deposits left at successively lower levels on the sides of the present river canyons as the streams cut downward to their present beds. The flows of lava and volcanic mud, which occurred in the Sierra Nevada during the Tertiary, and which preserved for the modern miner the ancient river channels, are lacking in this region. The terrace deposits of Siskiyou and adjacent counties date from Quaternary to Recent time. They parallel the present streams and have practically the same grades, and such earth movements as have occurred since their deposition have been probably slow and widespread, tending to rejuvenate the streams and accelerate their powers of erosion. The downward cutting rivers enriched their gravels by erosion of the ancient gold-bearing schists and the pre-existing

---

[1] U. S. Department of Agriculture, Weather Bureau, Summary of Climatological Data for the United States by Sections, Reprints of Sections 15 and 16. (Data are for periods earlier than 1923.)

[2] Averill, C. V., Preliminary Report on Economic Geology of the Shast Quadrangle, Cal. State Mineralogist's Report XXVII, chapter for January, 1931.

[3] Logan, C. A., Siskiyou County, Cal. State Mineralogist's Report XXI, chapter for October, 1925.

river and shore gravels of the 'Cretaceous Island' of northern California. This Cretaceous island embraced practically all of western Siskiyou County, this area having been elevated above the sea since very early ages, and being devoid of Cretaceous or later marine sedimentary rocks west of the vicinity of Cottonwood Creek. The chief tributaries which themselves have been the scene of placer mining and have contributed gold to the Klamath, are Scott River, Salmon River, tributaries of Shasta River, and Humbug Creek, all entering from the south, and Indian, Beaver and Cottonwood creeks from the north.

Scott River Region.

On Scott River early day placer mining camps flourished on the South Fork around Callahan and at Scott Bar, 3 to 4 miles from the mouth of the main stream. There is still considerable unworked gravel near Callahan. This is characterized by lack of grade and dump, and heavy boulders in places. The small area at Scott Bar was extremely rich. There remains quite an area of placer ground at the Roxbury Mine, and some high ground on both sides of Scott River, similar to the Quartz Hill diggings, where auriferous seams and veins of quartz are being worked by the hydraulic process. This seam belt has been considered responsible for the enrichment of the river below its crossing. It is noteworthy that from Callahan to Scott Bar the river itself has yielded little gold, but considerable profitable mining has been done in Quartz Valley, an arm of Scott Valley west of the present river. The gold on the west side of this valley was derived from an ancient channel, probably of Scott River, remnants of which are to be seen at Little Sniktaw Valley, and it can be traced southeastward upstream past Shackelford and Mill creeks, the Pinery workings (formerly Stockton Gravel Mining Company) and old Etna Mills. In places this channel has been entirely eroded into the valley, but elsewhere it is in place and has been worked on a small scale only, because of lack of water. It is a large channel with rather heavy, tight wash. Another class of deposit mined here was derived from local erosion of auriferous seams and small veins in the hills separating Quartz Valley from Oro Fino and Scott Valley. Quartz Valley has interesting possibilities for placer mining yet, but as the ground mentioned near its lower end has not been tested, it is not known whether it could be dredged, or would have to be worked by drifting. The ground is probably quite deep.

North Central Region.

Humbug district, northwest of Yreka and south of Klamath River; Cottonwood district, north of the Klamath, and quartz properties in the vicinities of Yreka, Fort Jones, Cherry Creek and other small quartz and pocket mining districts, may all be grouped together geologically. They occupy an area of 'Paleozoic Metamorphics' which includes very old rocks, the youngest of which may be Carboniferous and the oldest Cambrian, but which can not be definitely placed in the geologic column because of a lack of fossils. Both the igneous and originally sedimentary members of this 'catch-all' classification have been altered. Limestone, slate, some jasper, quartz porphyry, diorite, diabase, hornblende schists, and great masses of serpentine occur. The Paleozoic metamorphics have been compressed and show schistosity generally striking north to northeast and dipping east. Some of the principal veins follow this same strike and dip. There have been several subsequent minor movements, indicated by faults, shear zones and narrow veins and seams of quartz, striking in different directions. These in many cases faulted the earlier veins, as at the Hazel Mine. Many of the smaller pockety deposits are associated with intrusives such as diabase. In the Humbug district, some good ores were worked in serpentine, the principal mines having been the Boyle, Mountain Belle and Spencer. In general the work has been superficial. The best known mines in this entire region are the Morrison and Carlock, in Quartz Valley district, 4 miles northwest of Greenview, and the Hazel Mine southwest of Hornbrook.

Placer mining in this region has been on a small scale in later years, due to exhaustion of the more accessible ground and the protracted shortage of water. Developments in dredging are mentioned under the heading, Gardella Dredge.

From Cottonwood Creek southward along Shasta Valley, and past Yreka, extend the ancient sedimentary deposits flanking on the east what was formerly the so-called 'Cretaceous Island,'* previously mentioned. These beds consist of clay shales, sandstones, and shore gravels or conglomerates. The shale and sandstone contain beds of coal, and some of the beds are remarkably rich in shell remains and casts. This series in the region of the coal workings south of Ager, dips east 23° to 40°, the former figure being nearer the normal dip. These beds have been broken by a series of basaltic intrusions which show in knolls throughout the valley. The conglomerate was cut by Klamath River and its nearby small tributaries in the region of Hornbrook and Cottonwood Creek, and was extensively eroded. This conglomerate was gold bearing and contributed a part of the gold mined from the placers of Cottonwood, Rancheria, and other nearby creeks, and the immediate section of the river. Eastward of the conglomerate, the gravels of Klamath River are said to be barren. A very interesting retailed account of this gold-bearing conglomerate written by R. L. Dunn, appeared in the Twelfth Report of the State Mineralogist, pages 459–471. The placers of the Cottonwood region, including Rancheria Creek, are estimated to have yielded about $4,000,000. According to figures of yield reported by Dunn in the above paper, from workings then being mined, the conglomerate in the Blue Gravel Mine yielded 60 cents per square foot of bedrock when hydraulicked, this being only a part of the gold content, due to loss in the unbroken masses of conglomerate passing over the dump. The portion next the bedrock, when worked in an arrastra, paid 88 cents per square foot of bedrock. The conglomerate is overlain by Cretaceous sediments. The easterly area of sandstone and shale has been classified as Tertiary. Several unsuccessful attempts have been made to bring in oil wells in this vicinity and southward. Gas (probably dry marsh gas) was reported from a well 3½ miles south of Montague on the road to Grenada, and gas and alkaline water came from another shallow well. Six miles southeast of Montague, in the valley of Little Shasta River, some fresh water springs have been observed to emit gas.

* See State Mineralogist's Report XXVII, pp. 5–25, for general geology of this area.

All the north-central portion of the county, as described above, is accessible by roads from Yreka and other nearby points on the Pacific highway. Yreka, the county seat, is 286 miles north of Sacramento by paved highway. The entire region is served with electric power by California-Oregon Power Company. Timber is lacking in the Shasta Valley section, but is obtainable in the western part of the area. Elevation of this region ranges from 2000 feet to 5000 feet and there is some snowfall in all parts of it.

Klamath River District.

Until a few years ago the terrace deposits and bars along the Klamath River, and the placers along several of its tributaries, were the scene of the principal mining operations in the northwestern part of the county.

The country tributary to the Klamath on both sides remains for the most part difficult of access. The road following the river from the Pacific highway near Hornbrook to Martins Ferry climbs over the Bald Hills of Humboldt County to a connection at Orick with the coast highway. It is 76 miles from Hornbrook to Happy Camp over this road. For a distance of 120 miles from near Hornbrook to Somes Bar, it follows the course of the Klamath River in this county and makes accessible one of the most extensive mining regions of the state. While the smaller and more easily worked placers along the river have been mined, there remain many terrace gravel deposits of interest to the hydraulic miner and numerous low bars offering possibilities for small scale drift mining operations. The gold quartz veins have not yet been prospected far enough to permit a fair opinion of their possibilities, but some very promising showings are being made. The area from the river northward to the Oregon state line and westward to Del Norte County, comprising many separate isolated districts, may be classed as the base metal district. At the west end, in the Preston Peak country, are numerous small copper prospects, now all idle. The Buzzard Hill Mine on the south and the Grey Eagle property north of Happy Camp mark a valuable copper mining district, sufficient work having been done to prove large orebodies in the latter mine. Quicksilver ores occur along the watersheds of Beaver Creek and Empire Creek, and copper and zinc ores have been noticed in the latter locality. From Indian Creek eastward to the railroad a number of gold quartz mines, mostly of pockety nature, have been opened. Prospecting and development of such properties is now going on along the upper courses of Indian, Thompson, Horse, Beaver and other creeks. On Independence Creek, 14 miles south of Happy Camp, the Independence Mine has made a fine showing of phenomenally rich gold specimen ore.

The Klamath River district as a whole is well supplied with standing timber and water for mining purposes. While the people of California have forbidden the building of dams in the Klamath River, the question of utilizing waters of the river for hydro-electric power generation is complicated by the matter of jurisdiction, most of this mining region being in the national forest, and to a certain extent under federal control.

At Happy Camp local miners think that large yardages of gravel left behind by former hydraulic operators can now be worked at a profit with gold at $35 per ounce. A good water-right is said to be available on Elk Creek with the building of 12 miles of ditches and flume. This water could be used to work the Richardson, Knownothing, Collins, Reeve, Little Crumpton, Davis and other properties.

Salmon River District.

This district has been the most productive, and remains the most active and important gold mining area of the county. It embraces the drainage of Salmon River, an area of about 800 square miles, including many formerly productive quartz mines and placer deposits, the latter exploited in later years almost entirely by the hydraulic process. Considerable quartz prospecting is going on, and there is also a revival of hydraulic mining, which for several years had been hampered by shortage of water.

The principal mining camps of the district are Sawyers Bar, on the North Fork of Salmon River 55 miles by road from the railroad at Gazelle, and the same distance from Yreka, the county seat; and Forks of Salmon, at the junction of the north and south forks of Salmon River, 17 miles by road west of Sawyers Bar. Between Etna and Sawyers Bar, a distance of 25 miles, the road passes over the Salmon Mountains at an elevation of 5929 feet, and the heavy snowfall in normal years prevents travel by automobiles into the district during about four months of winter. A road along Salmon River from Forks of Salmon connects with the Klamath River road at Somes Bar, and gives this district an outlet to the coast system of highways via Orleans. The more important placer mining operations are confined to the benches, back channels and bars of the North Fork of Salmon River between Sawyers Bar and Forks of Salmon, and to Eddys Gulch, a tributary of north fork entering it from the south at Sawyers Bar. There are also unworked placer deposits in Whites Gulch, a tributary of north fork entering it three miles east of Sawyers Bar; on the south fork in the vicinity of Cecilville and Forks of Salmon, on the main river down stream from the latter camp and some small areas on the tributaries near the main branches of the river.

There are only two or three small areas of a few acres each of arable land along Salmon River, the balance of the district being an elevated mountain region, deeply dissected by the streams whose beds lie from 2000 to 4500 feet below the surrounding summits, so that the traveler is constantly climbing.

There is a good supply of timber for all mining purposes in this district. Water is available in the numerous creeks and gulches and in the Salmon River, in quantity ample for hydraulic and quartz mining, and in some places under conditions favoring hydro-electric power installations. The possession of these natural resources, and an exceptionally healthful climate, help to discount the remoteness of the district. The main North Fork of Salmon River has a grade of about 51 feet per mile from Sawyers Bar to Forks of Salmon, and the south fork 40 feet per mile.

The placers of the North Fork of Salmon River, in a distance of 17 miles between Sawyers Bar and Forks of Salmon have made an estimated production of $25,000,000 in gold. The sources of this were the numerous gold-bearing veins of the basin drained by Salmon River and its tributaries, most important of which were the last two named gulches, Russian Creek, and the south fork below the mouth of Black Bear Creek.

Quartz mining and prospecting have been very active in the Salmon River District, during 1934 and 1935, particularly in the vicinity of Sawyers Bar and Cecilville. Substantial production has been made by the King Solomon Mines Co. and Gold Ball Mining Co., and some production by Norcal Mining Co. and Mayland Mining Co. The Jumbo, Keaton, Hickey and other properties are being prospected. Further details on these mines are given below. Additional production will probably soon be made as a result of this activity. These mines are associated with a belt of slaty schists (closely similar to the Calaveras slates of the Mother Lode country) and other schists derived from both sedimentary and igneous rocks. Other igneous rocks, now considerably altered, intrude these schists.

Elliott Creek District.

This district is on the north side of the Siskiyou Mountains, the creek being a tributary of Applegate River. The principal property is the Blue Ledge Copper Mine, claims of wrich extend through the north-central part of T. 47 N., R. 11 W., and the south-central part of T. 48 N., R. 11 W. There are some other copper claims in the district. A number of the placer mines along Elliott Creek are described in State Mineralogist's Report XXVII,* copies of which are still available at offices of this division. Hence those descriptions will not be repeated here. The region is reached from Jacksonville, Oregon, 33 miles northeast.

Bullion Mountain District.

Prospecting is quite active in this district, and there is some small-scale production. In the properties visited, the veins follow one or both walls of basic dikes intruded into diorite. Mines include the Golden Rule, Lumgrey, Blue Jay, Trust Buster, Corbett, Bowser, Central and Iron Dike. Descriptions of the three last named are given below under 'Gold Mines'.

Other Minerals.

Among the minerals found in this county, and mentioned in this report besides gold, are asbestos, californite, chromite, coal, copper, lead, limestone, manganese, marble, molybdenite, platinum group metals, pumice, quicksilver, sandstone and miscellaneous stone. There are also numerous mineral springs. At present, production of other minerals is small as compared to the production of gold. A number of the gold mines and other deposits of Siskiyou County were described in State Mineralogist's Report XXVII,* chapter for January, 1931. Only the more active of these were visited again during the field-work for the present report. Descriptions of some of the deposits are repeated here to make the report more complete; also parts of descriptions of deposits of other minerals from State Mineralogist's Report XXI**, chapter for October, 1925, now out of print. Most of the gold mines described below were visited late in 1934, or early in 1935.

---

* Averill, C. V., Preliminary Report on Economic Geology of the Shasta Quadrangle, Mining in California, January, 1931, pp. 56, 57.
** Logan, C. A., op. cit.

MINERAL PRODUCTION OF

Year	Gold, value	Silver, value	Chromite		Mineral water	
			Tons	Value	Gallons	Value
1880	$440,735	$95,340				
1881	850,000	1,500				
1882	720,000					
1883	400,000					
1884	475,000					
1885	338,659					
1886	342,677	64				
1887	606,859	177				
1888	625,000					
1889	915,294	370				
1890	860,303	23				
1891	957,220	120				
1892	1,013,332	56				
1893	799,108					
1894	760,782					
1895	950,006	177			200,000	$80,800
1896	1,091,265	653			ª	
1897	842,123	34			ª	
1898	768,804	321			ª	
1899	991,771	100			ª	
1900	951,397	ª6,700			700,000	45,000
1901	886,043	ª2,980			700,000	175,000
1902	906,989	233			750,000	187,500
1903	613,576	22			750,000	50,000
1904	892,685	1,230			750,000	50,000
1905	803,035	2,499			ª	
1906	ª	ª			ª	
1907	398,017	3,037			725,000	36,250
1908	504,156	6,125			700,000	80,000
1909	416,160	2,145			500,000	10,000
1910	437,376	2,322			500,000	60,000
1911	422,297	2,561			700,000	120,000
1912	472,314	2,980	220	$2,310	700,000	120,000
1913	ª180,125	ª1,228			700,000	120,000
1914	312,842	1,026			650,000	65,000
1915	426,716	2,081	ª		626,680	62,990
1916	441,307	2,312	2,251	28,731	502,650	50,530
1917	325,550	16,883	2,046	49,797	503,000	50,600
1918	294,227	14,501	6,612	336,588	501,750	50,175
1919	226,525	17,049	510	13,379	451,500	90,375
1920	80,707	5,218	215	5,732	300,150	60,015
1921	42,635	294	ª		250,150	5,015
1922	75,105	612				
1923	45,633	298			200,150	4,042
1924	63,570	296				6,100

SISKIYOU COUNTY, 1880-1934

Platinum group metals		Miscellaneous stone, value	Miscellaneous and unapportioned		
Ounces	Value		Amount	Value	Substance
100	$600				
				$1,202,742	Unapportioned, 1900-1909
			200 lbs.	23	Copper
1.6	21				
5.3	93		2,500 cu. ft.	1,250	Sandstone.
			2,500 cu. ft.	1,500	Sandstone.
			193 lbs.	39	Copper.
			2,643 lbs.	140	Lead.
		$39,000	11,433 cu.ft.	12,897	Sandstone.
			1,000 bbls.	1,000	Lime.
			220 tons	300	Limestone.
			4,949 lbs.	1,183	Lead.
			1,800 cu. ft.	1,485	Sandstone.
			1,090 lbs.	1,680	Lime.
			3,360 lbs.	144	Lead.
			50 tons	500	Pumice.
		5,028	1,050 cu. ft.	1,750	Sandstone.
			100 bbls.	300	Lime.
			2,225 tons	2,200	Limestone.
				14,745	Gems.
			1,204 cu. ft.	2,000	Sandstone.
		9,475	335 bbls.	735	Lime.
			35 tons	525	Limestone.
				1,000	Gems.
			150 bbls.	120	Lime.
		6,580	24 tons	24	Limestone.
			650 cu. ft.	455	Sandstone.
			250 cu. ft.	250	Sandstone.
		609		250	Gems.
				250	Gems.
		4,883	90 tons	2,000	Pumice.
				1,500	Other minerals.
			100 tons	500	Coal.
9	304	5,371	58 lbs.	2	Lead.
			677 bbls.	629	Lime.
			250 cu. ft.	150	Sandstone.
			188 lbs.	9	Lead.
		4,630	745 bbls.	745	Lime.
				16,923	Chromite, copper, marble, sandstone.
1		45,407		12,609	Copper, building stone, lime, platinum, sandstone
				500	Granite.
15	709	134,382	888,043 lbs.	242,436	Copper.
			192 lbs.	17	Lead.
				8,535	Lime, sandstone, soda.
1	58	24,588	573,593 lbs.	141,677	Copper.
				15,473	Lead and pumice.
7	1,015	26,405		111,294	Copper, limestone, pumice, quicksilver.
		30,322		47,121	Copper, lime, limestone, potash, pumice, quicksilver.
1		44,343		1,060	Asbestos, brick, chromite, lime, platinum.
		21,726		4,020	Other minerals.[5]
3	339	129,291		1,408	Other minerals.[6]
		67,787		3,034	Other minerals.[7]

Year	Gold, value	Silver, value	Chromite		Mineral water	
			Tons	Value	Gallons	Value
1925	$180,120	$831			³	
1926	141,240	709			³	
1927	138,822	586			³	
1928	85,717	421			³	
1929	63,843	863			³	
1930	70,332	4,172			³	
1931	74,326	169			³	
1932	133,115	304			¹	
1933	324,954	686			¹	
1934	528,395	1,861			¹	
Totals	$26,708,789	$204,169	11,854	$436,537	¹12,361,030	$1,579,392

¹ Includes crushed rock, rubble, rip-rap, sand, gravel.
² Recalculated to 'commercial,' from 'coining value' as originally published.
³ See under 'Unapportioned.'
⁴ Production from dredging operations included in Stanislaus County production.

ASBESTOS

There are many large areas of serpentine, in which some prospects of asbestos occur and others may be expected, in this county. Most of these are too far from railroad for present consideration. Among the remote, unprospected areas may be mentioned the dikes and larger bodies of the rock which occur in the western part of the county, beginning six miles north of Orleans, thence recurring frequently along both sides of Klamath River for many miles, and along the west side of the county to the Oregon line, as well as eastward along Salmon River from Somes Bar at intervals as far as Methodist Creek; near Hamburg Bar, north of Klamath River; near Scott Bar and in the Callahan district. Among the serpentine areas within reasonable distance of the railroad are the mountainous area adjacent to the highway between Yreka and Fort Jones, and areas near Yreka, near Montague in Shasta Valley, and near Gazelle and Dunsmuir.

Burns Ranch. Owner, W. L. Burns, Gazelle. Patented land in Sec. 16, T. 42 N., R. 6 W., four miles west of Gazelle.

A little shallow work has been done on a chrysotile asbestos prospect, showing fiber up to 1¼ inches in length, but brittle. The quality might improve with depth.

C. C. Cady Asbestos Prospect on Greenhorn Mountain, between Yreka and Fort Jones, was leased in 1923 to Geo. Souza who did some work. The surface showing of asbestos here can be traced for a considerable distance.

Geo. W. Conrod, P. O. Box 577, Weed, sent in to the Bureau late in 1925, samples of slip-fiber (amphibole) asbestos. Locality not stated.

H. Johnson of Sisson has seven locations for asbestos on Eddy Mountain, southwest of Weed and near the Trinity County line.

J. J. Murray of Yreka several years ago located asbestos claims near the head of Seiad Creek, on a steep mountain four to five miles

SISKIYOU COUNTY, 1880-1934—Continued

Platinum group metals		Miscellaneous stone[1], value	Miscellaneous and unapportioned		
Ounces	Value		Amount	Value	Substance
				$3,535	Lime and limestone.
		$23,800		11,340	Mineral water, platinum, sandstone.
16	$1,780	327,569		22,853	Coal, lead, mineral water, sandstone.
10	690	102,428		56,420	Mineral water, sandstone.
		370,833		14,195	Copper, lead, gems (rhodonite), mineral water.
		110,878		54,205	Copper, lead, limestone, quicksilver, mineral water.
		85,851		75,046	Copper, lead, granite, mineral water, gems, platinum, quicksilver, lime, pumice.
		79,772		32,740	Other minerals.
		23,415		27,185	Lead, quicksilver, mineral water.
		29,036		19,502	Copper, lead, mineral water, pumice.
		67,216		50,694	Copper, lead, mineral water, pumice, tube mill pebbles.
167.9	$5,609	$1,780,625		$2,228,844	

[4] Includes limestone and mineral water.
[6] Includes lead and lime.
[7] Includes coal, limestone, lime and platinum.

from the end of the wagon road, north of Seiad P. O. and nearly 70 miles from the railroad. Idle.

Shasta View Asbestos Prospect. Owner, W. S. Russell, Edgewood. In Sec. 8, T. 41 N., R. 5 W. Contains 12 unpatented claims, reached by three miles of good road from Edgewood. Elevation 4000 feet. There is plenty timber and water. Electric power is three miles distant.

Chrysotile asbestos occurs here and in 1921 a production of a small quantity of short fiber chrysotile was made. It was stated this carried 25% of fiber. Samples from the property show fully that percentage of fiber, one-half inch and less in length. There is some fiber one inch long.

National Cement Company of Modesto did some work in 1921–22 on the property, but it later reverted to Russell and is idle, except for assessment work. Development consists of short open cuts, not over 16 feet deep.

CHROMITE

Considerable chromite was produced in this county during the war, and some of the remoter districts were just getting into production when the armistice was signed.

At the height of the war minerals 'boom' in 1918 the State Mining Bureau published a book of 248 pages, designated as Bulletin 76, "Manganese and Chromium in California," in which were listed and described practically all the then known deposits of these minerals in the state, and this book is still available at fifty cents a copy. As there has been no market worth mentioning for chromite from this state since 1918, except small lots of high-grade ore from properties nearby the railroad, there appears to be no occasion for extensive field work in this connection. The main facts regarding chromite properties in the county are given in the synopsis following.

2—25888

Attention should be called to the fact that since 1918 the Klamath River road has been made passable for trucks the entire distance from Hornbrook on the railroad to Somes Bar in the southwest corner of the county, where connection is made with the coast highways via Orleans. This opens up the large and only partly prospected areas of serpentine near the road, and is an outlet for the remoter areas west and northwest of the river to Del Norte County on the west and the Oregon state line on the north. This district was promising to be a big producer of chromite when the market failed, but the cost of hauling would be very heavy, as Happy Camp is 76 miles from Hornbrook.

The districts from which chromite production has been made in the county are: Callahan, Etna Mills, Fort Jones, Scott Bar, Yreka, Gazelle, Hamburg Bar, and Dunsmuir. In addition to these, and the others mentioned previously, there are other serpentine and peridotite areas (listed under Asbestos) not so far productive.

The production of individual properties was in few cases over 500 tons, the largest producer being the Coggins deposit, three miles south of west from Dunsmuir, where the ore contained 38% to 40% Cr_2O_3. The highest grade ore came from the Callahan district, where claims were 30 to 50 miles from Gazelle, the nearest shipping point. A large property near Hamburg Bar is about 48 miles from Hornbrook, and as it lies on the north side of Klamath River would require a bridge or tramway.

Besides the above localities, the following deposits have come to our attention since 1918, in addition to those listed in the tabulation of properties taken from our Bulletin 76:

Chromite King—see Peg Leg.

Furlong Bros. and T. A. Lloyd have been prospecting lenses of chromite with small open cuts in T. 41 N., R. 7 W., one to three miles south of the point where the road from Gazelle to Callahan crosses the summit.

H. Johnson of Sisson reports having a large tonnage of chromite six miles from the railroad, and within two miles of an old road.

Milne and Reichman Deposit. Owners, Geo. Milne and G. A. Reichman, Fort Jones. This is a large disseminated deposit of chromite 1½ miles from the Scott Bar road, 44 miles west of Yreka. The chromite occurs in seams and high grade ore must be sorted. The average is low grade. Six hundred tons have been mined, and most of it hauled to Yreka. Idle since 1919.

Mountain Chrome Mine. Owners, W. D. and N. Dale, address W. D. Dale, 2668 Fourth Avenue, Sacramento. This is seven miles southwest of Edgewood, in Sec. 24, T. 41 N., R. 6 W., near North Fork of Shasta River. The owners state that 100 tons of ore have been mined and that they have crosscut the ledge for 80 feet without striking the other wall.

Peg Leg Mine. Former owners, W. P. Johnson, N. Lambert and Mrs. F. A. Shebley. Address N. Lambert, Fort Jones. This is 14 miles southwest of Yreka via road to Fort Jones, and near Moffit Creek. High grade chromite, containing 50% Cr_2O_3 was mined in 1918 and

1919. Ninety tons was shipped, an equal amount was on hand and the prospects were good for opening a larger body in the lower tunnel when visited by a Bureau representative in July, 1919.

Walter E. Brown and Major H. A. White of Yreka are holding seven claims to cover this deposit. A few tons of high-grade chromite were piled on the dump at the lower adit, which was partly caved when the property was visited in July, 1935. The name has been changed to *Chromite King*.

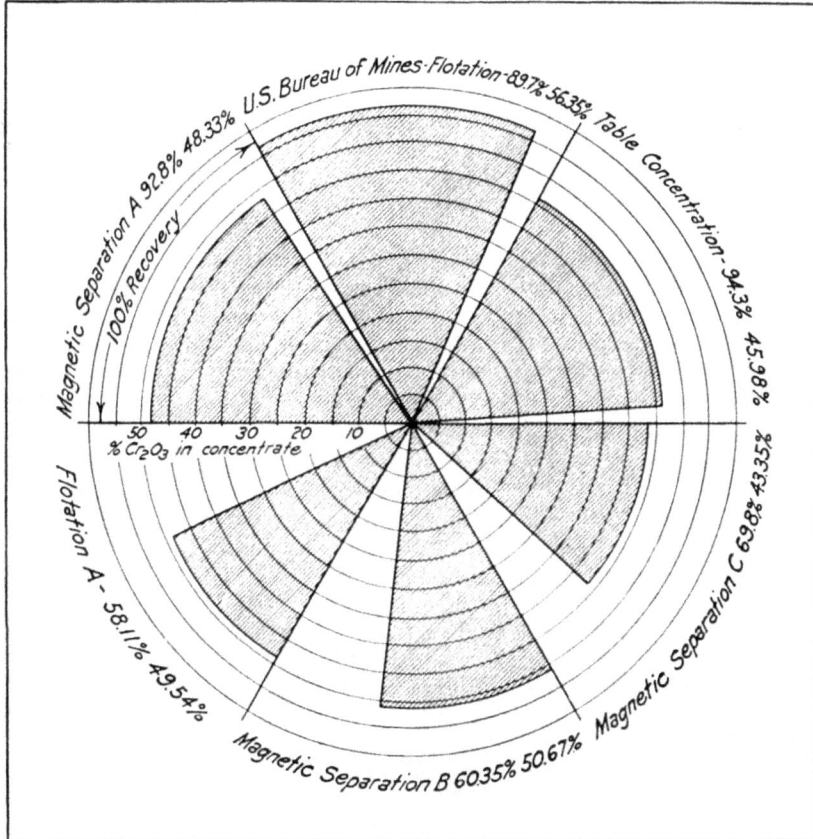

Fig. 1. Seiad Creek Chrome Deposits. Results of metallurgical tests made for H. W. Gould and J. H. Hines, in percentage of recovery and percentage of Cr_2O_3 in concentrate, by six different methods.

Seiad Creek Chrome Deposit in Sec. 20, T. 47 N, R. 11 W., is held by H. W. Gould, Mills Building, San Francisco, and is under option to J. H. Hines, Goodwyn Institute Building, Memphis, Tennessee. The total distance from the railroad at Hornbrook is 58 miles, 27 miles paved and oiled road down Klamath River, then 24 miles gravel and dirt road to Seiad, then seven miles mountain road up Seiad Creek, and 1000 ft. steep sled-road to mine. The deposit is on a ridge between two forks of Seiad Creek, each of which flows a good stream of water the year round; and the country is well timbered.

LeRoy A. Davis of Seiad, and Cyrus B. Crawford originally located three claims (about 1916), and Gould added four additional locations in 1933. During the war several thousand tons of hand-sorted chromite running 48% were shipped, part of it by the late Dr. Reddy of Medford, Oregon, who held a lease on the property.

The deposit is more vein-like than the lenses of chromite usually found. The chromite occurs as veins and veinlets following the schistosity of a schist (or gneiss) apparently formed from a basic igneous rock. An occasional vein is 2 ft. to 3 ft. wide, and has the lenticular form commonly found. In addition to these there are thousands of tons of schist containing veins and veinlets of chromite varying from a fraction of an inch to an inch in width. Gould has made estimates of 1,000,000 tons to 2,000,000 tons of ore that will concentrate to 500,000 tons of product assaying 50% Cr_2O_3, assuming a minable width of 35 ft., length 1800 ft. to 2000 ft. and depth 250 ft. These figures give a good idea of the appearance of the deposit; but additional development work will be required to actually prove the tonnages mentioned.

Development work consists of some 30 surface cuts, the largest of which, 125 ft. by 40 ft., produced most of the ore shipped during the war, and 600 ft. of underground work in two adits. The latter have been driven into both sides of the ridge between the two forks of the creek, and comprise 200 ft. of crosscutting and 400 ft. of drifting. The following analysis of the chromite was made for Gould by Abbot A. Hanks of San Francisco:

Cr_2O_3	46.39%
FeO	16.87%
SiO_2	8.54%
Al_2O_3	10.30%
MgO	16.22%
CaO	1.40%

This indicates a ratio of chromium to iron that is higher than usual, which is an advantage in making a high-grade concentrate.

The accompanying diagram (Fig. 1) shows the results of experimental work done on the metallurgy, which is described in greater detail in a report by Gordon I. Gould, dated Feb. 20, 1934. In both table concentration and magnetic concentration, difficulty is experienced with the fine sizes that are unavoidably made in crushing and grinding. Good results can be obtained by both methods on sized material of about 40-mesh size; but results are poor on the 50% of finer sizes that would be made. Much of the gangue is olivine, which approaches the specific gravity of chromite more closely than other gangues. On a concentrating table, the finer sizes of this are trapped in the concentrate. Three separate and distinct types of magnetic separator were used with the results indicated in the diagram. The magnetic permeability of chromite (particularly this chromite with its low iron content) is low, and the gangue has a slight permeability. In the finest sizes, this effect together with molecular attraction and dusting makes separation difficult. Best results were obtained by the Mississippi Valley Experiment Station of the U. S. Bureau of Mines, Rolla, Missouri, using floatation. The only details available on the

method used are given below: An acid circuit was used on account of
the olivine-gangue. Ore tested assayed 32.17% chromite (Cr_2O_3).

Product	Weight (%)	Assay per cent Cr_2O_3	Per cent of the Cr_2O_3
Concentrate	46.3	57.76	83.1
Middling	13.2	22.88	9.4
Tailing	40.5	5.95	7.5

In the above test rougher concentrate assaying 50.02% Cr_2O_3 was
cleaned twice, but one cleaner produced concentrate that assayed
56.35% Cr_2O_3 that represented a recovery of 89.7%. The chromite
was floated in this test. Other tests were made on floating the gangue
with good results, but not as good as those given above.

Several mining claims on a chromite deposit in Sec. 30, T. 45 N.,
R. 10 W., on McGuffy Creek, a tributary of Scott River, are held by
H. W. Gould also. They are under option to J. H. Hines.

Other properties mentioned in Bulletin 76 are tabulated below:

Name of property	Section, township and range	Length of haul, railroad shipping point
Ball Ranch	16–41– 9	26 miles Yreka
Burns Ranch	16–42– 6	4 miles Gazelle
Butcher Ranch	---------	1 mile Yreka
Coggins	---------	3 miles Dunsmuir
Chastain	---------	---------------------
Cramer	22–44– 8	14 miles Yreka
Hamburg Bar	{15–46–11 / 22–46–11 / 23–46–11}	48 miles Hornbrook
Davis	10–39– 9	38 miles Gazelle
Dexter Ranch	---------	4 miles Montague
Facey	---------	---------------------
Flederman	10–41– 9	28 miles Yreka
Flederman Lease	--43– 8	15 miles Yreka
Grant Lease	25–42– 9	14 miles Yreka
Grouse Creek	{13–40– 8 / 19–40– 7}	25 miles Gazelle
Le May	--44– 7	---------------------
Martin-McKeen	34–40– 9	36 miles Gazelle
Masterson	14–40– 8	Gazelle
McCarthy	---------	32 and 42 miles Yreka
Musgrave	---------	---------------------
Souza Ranch	---------	1 mile Yreka
Sugar Creek	---------	40 miles Gazelle
The Chrome Mine	12–39– 9	36 miles Gazelle
Valine Ranch	---------	1½ miles Yreka
Wilson Ranch	---------	20 miles Yreka
Wurst	---------	35 miles Gazelle

At the time the chromite market collapsed, Siskiyou County was
regarded as one of the largest prospective producers of the mineral
in the United States, and it was estimated that the production in the
next year would be 30,000 tons or more, mostly from the Klamath
River deposits, located from Hamburg Bar westward.

COAL

Lignite and sub-bituminous coal occur in Siskiyou County in Shasta
Valley. Probably several thousand acres in this valley are underlain
by coal.

Ager Coal Mine is on the Hagedorn Ranch, five miles south of Ager. Henry Hagedorn, owner. Years ago a slope 700 feet long on an incline of about 15° was sunk on the coal, which occurs between layers of hard shale, with alternating beds of hard sandstone above and below. The area has been mapped by Professor J. P. Smith* as Tertiary. The normal dip of the strata is about 23° east, but in this vicinity the sedimentary beds have been broken by a series of basaltic intrusions, indicated by a line of low hills running in a southeasterly direction. The dip of the beds increases to 40° NE. near this line. From the incline three drifts were run north a maximum distance of 500 feet, and three drifts south, the longest 400 feet, on the coal. A width of six feet carries coal, but the best, solid part of coal varies from 14 inches up to four feet, and is said to average two feet. The walls stood well without timbering. Five thousand tons or more have been produced and some of it sold. The last active work was in 1914–15 by Yreka Development Company, then lessee.

Numerous prospect holes have been drilled in the same vicinity on the *Hagedorn, Herr, Cooley* and *Denny Ranches*.

On the *Cooley Ranch* a hole 130 feet deep is said to have struck 11 feet of coal at a depth of 118 feet. Another hole 106 feet deep is said to have struck 20 inches of coal at a depth of 95 feet.

Siskiyou Coal and Coke Company. R. D. Clark, President, Yreka. Leverett Davis, Mining Engineer. This company had leases on parts of ranches in Shasta Valley where coal has been found, as mentioned. In September, 1925, the company was putting down a diamond drill hole on the Herr Ranch, one-fourth mile north of the 106-foot hole mentioned above. The hole was down 110 feet and had not struck coal, which was not expected until a depth of over 200 feet. The core of the hole at time of visit showed alternating layers of sandstone and shale. The sandstone was most abundant, and some of it is quite hard. The coal dips NE. 40° according to C. B. Kay, the steeper dip being on account of the proximity to the line of basalt intrusives. The company reports tests on the coal where previously opened indicating coking possibilities, but a classification as bituminous coal.

In the southeastern part of the county, in townships 39 and 40 north, ranges 1 west and 1 east, there occur outcrops of sedimentary rocks, forming the northern part of an area of Triassic and Tertiary sediments which extend north and east from near Redding. Just south of the Siskiyou County line in Shasta County some coal prospects have been noted in this area in Tertiary formations, but are not known to extend into Siskiyou County.

There is a possibility of coal being found at the proper horizon anywhere in the Tertiary beds which are exposed from the Oregon line southward to five miles south of Ager, having a maximum width of eight or nine miles in townships 47 and 48 north, ranges 5 and 6 west, M. D. M. Such an occurrence is found on the James O. McMaster Ranch in the Hornbrook district, where nine feet of shale is intershot with small streaks and spots of soft coal, and there is also a seam up to two inches thick of what McMaster describes as "good hard coal, making a hot fire with a blue-green gaseous flame."

* Cal. State Min. Bur. Geologic Map of Cal., 1916.

COPPER

Very little progress has been made in the county in copper mining the past 10 years. After the price of copper fell, following the war, there was little incentive for the copper miners of Siskiyou County to produce ore, and since 1920 practically only necessary assessment work has been done. Little field work was done, therefore, among the copper mines and prospects for this report, and the following notes are a compilation of past reports, scattered among various publications in the past, but gathered here for readier reference; also a list of newer prospects.

Siskiyou County apparently contains some of the best unworked copper deposits in the state. The two most promising districts are near Happy Camp in the western part of the county, and near the northern county line on the headwaters of Applegate River. The two principal proved properties are the Grey Eagle near Happy Camp, and the Blue Ledge in the Elliott Creek district. There are numerous copper prospects throughout both the western and northern parts of the county, where gossan outcrops occur and have been slightly prospected, but not sufficiently to definitely indicate their possibilities.

The greatest hindrance to development of this branch of mining is the remoteness of the region from railroads. Happy Camp is 76 miles from railroad and most of the copper prospects are still farther away. From time to time there is talk of a railroad being built from the vicinity of Hornbrook down the Klamath River. Due to the districts being in the national forest the federal government would not permit operation of a smelter.

Bulletin 50 of the California State Mining Bureau lists 45 prospects of copper in this county. This bulletin was published in 1908. Since that time, the Blue Ledge and Grey Eagle properties have been extensively developed and proved to contain considerable ore. Little has been done with the others. The principal production of copper occurred in 1917 and 1918, when about 1,460,000 pounds were sold. Developments in the past few years have served to emphasize the widespread distribution of copper prospects.

Ames Prospect. Owner, A. Ames, Happy Camp. One claim on a gossan outcrop on the southwest slope of Clear Creek Canyon, eight miles by road and three miles by trail from Happy Camp. Only assessment work is being done.

Barton Claims. Owner, H. J. Barton, Yreka. Claims on Horse Creek, nine miles upstream from the mouth. Formerly known as the Hetchell prospect. Contains four claims. Idle.

Barnum Brothers Copper Prospect. (Copper Mountain Group.) Owners, Barnum Bros. et al. This group comprises 19 claims on Sambo Creek in Oak Bar district. It was equipped with a compressor and air drills in 1919 and in 1925 was under lease to Williams of Yreka, who was continuing an adit. Several hundred feet of underground work has been done.

Blue Ledge Mine. Owner, Towne Mines, Inc., 120 Broadway, New York City. This property comprises 26 patented claims in T. 47 and 48 N., R. 11 W., in Elliott Creek district, at an elevation of about 5000 feet. Jacksonville, Oregon, is the nearest town and is 33 miles northeast.

The footwall of deposit is grey to black micaceous schist and the hanging wall is a white mica schist. The vein strikes nearly north, dips 60° west and averages 5 feet wide, though reaching a maximum width of 35 feet. A pay shoot 1500 feet long and 5 feet wide is reported proven by a series of adits and drifts which have developed the vein for a length of 1800 feet and a depth of 600 feet. In 1920 the management estimated that there were 240,000 tons of ore blocked out, having an average content of 4.8% copper, .035 oz. gold and 1.5 oz. silver. It is essentially chalcopyrite, and pyrite, the iron content going as high as 32%. The mine was operated up to the latter part of 1920 but has been idle since. Nearly 9000 tons of sorted ore was shipped. It contained an average of 13.7% copper, 5.5 oz. silver, .1 oz. gold, 30% iron and 30% sulphur. Steam was used for power, wood fuel from the claims being utilized. There are a compressor, air drills and underground hoist.

Bibl: Cal. State Min. Bur. Bull. 50, p. 123. R. XIV, p. 817; XVII, p. 530.

Buzzard Mine. See under gold quartz mines. The ore exposed in the mine workings immediately under the oxidized ore is a heavy sulphide, mostly pyrite, with a small amount of copper.

Clear Creek Mine. (*Davis Mine.*) Owners, Estate of Reeves Davis et al. It is on Clear Creek, 13 miles southwest of Happy Camp. It has been quite extensively prospected, and gave an encouraging showing, but it is said that a good sized orebody could not be found in place. *See Paradise*, under Gold.

Dillon Creek Prospects. In the summer of 1917, following a forest fire which cleared off the brush, eleven copper claims were located in the vicinity of the headwaters of Dillon Creek, which enters the Klamath from the northwest near Cottage Grove. Later the Grey Eagle Copper Co. did some prospecting in this vicinity, but the result of this is not known. The original locators of the claims were: Jettie Albars, Jack Davis, Swaney Peters, Wm. Elliott, Chas. and M. Thomas; Henry, Mrs. J. C. and Mrs. J. Aubrey, all residents of the Blue Nose and Somes Bar districts.

Efman & Boorse Prospect. Owners, Mike Efman and H. G. Boorse. Happy Camp. This is an undeveloped claim on a gossan outcrop in Elk Creek Canyon, two miles south of Happy Camp. Only assessment work was being done in 1925.

Gray Eagle Mine. Owner, Gray Eagle Copper Co., 14 Wall St., New York City. It comprises 32 claims, mostly patented, in Sec. 4 et al., T. 17 N., R. 6 E., H. M., on mountain on the east side of Indian Creek, north of Happy Camp and 80 miles by road west of Hornbrook, the nearest railroad station. The elevation of main adit, No. 7, is 2539 feet.

The deposit is an ore zone trending southeast, varying in width from 10 to 80 feet, and carrying ore along fracture planes normal to the grey schist country rock. The ore is principally chalcopyrite, carrying an average of 6% copper. The sulphide ore is covered by a gossan cap. No. 7 adit has been run about 3000 feet southeasterly and a total of about 10,000 feet of underground work is reported from this adit. Ore exposed on this level is said to be mostly pyrite, but where raises have been put up good ore has been exposed, as was also the case in the upper adits. Over 1,000,000 tons of ore are claimed to have been blocked out. The mine is equipped with a steam plant, large compressor, electric light plant, shops and buildings. Water is taken from nearby branches of Indian Creek. There were 60 men employed between 1916 and 1918, since when the mine has been idle.

Bibl: Cal. State Min. Bur. Bull. 50, p. 132. R. XIV, p. 818; XVII. pp. 531–532.

Henry Wood, Seiad P. O., has a copper prospect on Beat and Bender Gulch, near the Portuguese Mine, and has been doing some work the past summer.

Isabella Copper Mine. Owner, Isabella Copper Mining Co., c/o Loren W. Smith, 611 Bank of America Bldg., Oakland. This is a prospect containing 120 acres, in Sec. 34, T. 41 N., R. 7 W., 18 miles southwest of Gazelle. The vein is said to carry 'grey copper,' silver and gold. It has been developed by a shallow shaft and several adits. Equipment includes a steam plant, small air compressor and air drill.

Bibl: Cal. State Min. Bur. R. XVII, p. 532; XXVII, p. 28.

Liberty Bond Group. Owner, F. B. McCann, Happy Camp. This comprises 11 claims on a ledge in schist near a limestone contact. The ore is said to carry 1 to 1½% copper and several dollars gold per ton. Only assessment work has been done.

Bibl: Cal. State Min. Bur. P. R. 8, p. 14.

Malone Mine. Owners, Churchill, Roseburg et al. of Yreka. It is in Elk Creek Canyon, 14 miles by trail south of Happy Camp. Some good looking sulphide ore, carrying pyrite and chalcopyrite, has been exhibited from this property. The principal work is an adit 160 feet long.

Parker Group. Owner, Geo. J. Parker, Copper P. O. This comprises six unpatented claims in Sec. 20 or 21, T. 48 N., R. 11 W., in Elliott Creek district. About 200 feet of prospect adits have been run. The owner has been working alone.

Sunshine Group. Owner, W. J. Brown, Happy Camp. It contains two unpatented claims on a copper-bearing outcrop on Cave Mountain, five miles northeast of Happy Camp, three and one-half miles being by trail up the canyon of the creek. An adit has been run 130 feet at an elevation of 2040 feet, and it is expected the vein will be cut by raising 15 feet. Only assessment work is carried on.

Yellow Butte Mine comprising the W. ½ Sec. 25, T. 43 N., R. 4 W., is assessed to C. A. Stephens of Weed. It is 12 miles northeast of Weed by a good road. The deposit consists of quartz veins in diorite, which belongs to the older formations, and apparently formed a mountain, around which flowed the predominant lava of the region. It thus formed an island in these flows. An incline-shaft, said to be about 300 ft. deep, has recently been retimbered for 35 or 40 ft. The dip is 70°. The white quartz is heavily stained by green copper minerals. A section across the caved ground at the collar of the shaft appears as follows: diorite foot-wall, 6 inches quartz, 3 ft. diorite, 4 inches quartz, 8 ft. diorite, 2 inches quartz, diorite hanging wall. Sorted ore on the dump shows plentiful chalcopyrite and molybdenite; but this kind of ore was not seen in place. Apparently it came from some depth in a part of the shaft not yet cleaned out.

DIATOMACEOUS EARTH

Ed Ewing, Yreka, has written the Bureau that he has 520 acres of land on which diatomaceous earth outcrops. This deposit has not been visited by a Bureau representative, so no particulars are available as to its location or quality.

GOLD MINES

Balfrey Mines comprise 800 acres of patented land, NE. ¼ Sec. 29 and all of Sec. 28, T. 41 N., R. 7 W., 20 miles southwest of Gazelle. They are owned by M. H. Balfrey of Gazelle. A part of this property was described under the name, "Siskiyou Lead Mine," in State Mineralogist's XXVII.* According to Balfrey, there is a silver prospect on this land, also a vein carrying chalcopyrite and gold that has been developed with a 52-ft. shaft.

Big Cliff (Winterings Mine) is a property of 12 unpatented claims in T. 40 N., R. 10 W., 3 miles southeast of Findleys Camp. The claims are owned by Mrs. Margaret Wintering and Mrs. F. H. Osgood, and are optioned to Dr. W. W. Barham of Yreka. Wintering drove 2000 ft. of tunnels, and milled small amounts of high-grade gold ore. The whole mass of the siliceous schist here is said to carry low-grade values in gold; and it is thought that a large enough tonnage of this ore can be developed to pay by working on a large scale. The property has not been visited.

Black Bear Quartz Mine is assessed to R. S. Phippeny, Sawyers Bar. It comprises 70 acres, patented, seven miles by road southwest of Sawyers Bar in T. 39 N., R. 11 W. The production record has been the best for quartz mines of Siskiyou County, about $3,100,000. This mine was visited in 1933 and in August, 1934, but was idle at both times. For references to descriptions of earlier operations, see accompanying table of mines.

Blue Bar Placer comprises 30 acres in Sec. 31, T. 18 N., R. 7 E., held by C. Scott Greening of Happy Camp. It covers about half a mile of bars along the present channel of Indian Creek just below the Classic Hill mine. According to Greening, the gravel averages 15 ft. deep and 200 ft. wide. He says that the flow of water is so great that only a few feet near the surface could be mined in the early days, and that the gravel below is very high in grade.

Blue Eagle and *Black Dike Groups* of 28 unpatented mining claims in Sec. 10, 11, T. 39 N., R. 10 W., on the steep ridge between Six-Mile and Trail Creeks, at an elevation of 6500 to 7000 ft., are held by H. D. Winship and J. W. Preston, 350 Post Street, San Francisco, Geo. F. Bartlett, Bob McLane, T. T. Taylor and Cutler Paige. From the railroad station at Gazelle, the property is 42 miles southwest. Gazelle to Callahan, 28 miles dirt road, then six miles mountain road up the South Fork of Scott River, then eight miles mountain trail. Roughly

_____
* Averill, C. V., *op. cit.*, pp. 60, 61.

200 ft. of adits were run here 60 or 70 years ago by King and Ault, and some production of gold was made.

The deposit consists of quartz veins and veinlets in schist of several different kinds.   On the ridge above the workings is a graphitic schist similar to that at Sawyers Bar.   At the workings is a contact between a green, chloritic schist and a siliceous schist, in part replaced by white quartz.   Nearby is an igneous intrusion (dioritic?).   An adit, 80 ft. long, was run years ago on a stringer zone on the contact between the chloritic schist and the siliceous schist.   Recent work in this adit consists of an 18-ft. winze, in which Bartlett states that the quartz stringer zone assays $27 in gold across a width of $3\frac{1}{2}$ ft.   This adit is 700 ft. lower in elevation than the top of the ridge, where some open cuts have been made to expose the schist and stringers in place.   A lower adit has been run; but it apparently is entirely in the siliceous schist footwall formation.   Prospecting is planned to determine whether the entire mass of the schist contains enough gold, including the quartz veins and stringers, to pay for mining on a large scale.   This is complicated by a heavy overburden, 4 ft. to 5 ft., of soil and loose broken rock; and it is probable that prospecting will take the form of following the $3\frac{1}{2}$-ft. zone of quartz stringers.   Then cross-cuts can be run to explore the schist.   One man is prospecting the property.

Blue Gravel Claims (Placer), of 100 acres, are in Sec. 35, 36, T. 17 N., R. 7 E. and Sec. 1, 2, T. 16 N., R. 7 E., and are held by Frank Murree and John Whittaker, Happy Camp.   They lie between Cade Creek and Ranch Creek, which flows into the Klamath River at the Reeve Ranch.   Prospect cuts show the presence of a terrace here, some 300 ft. higher than the present river.   The channel extends through the Crumpton, 200 acres, patented, and the Whittaker, 80 acres, patented, in Sec. 2, adjoining.   On the Crumpton, several acres have been mined off, and banks 40 and 50 ft. high remain, from which gold production is said to have been $25,000.   A 10-mile ditch from Indian Creek, with outlet 700 ft. higher than Klamath River, has been surveyed to work these properties; but construction has not yet started.

Blue Horn Mine of two unpatented mining claims is owned by R. D. Freshour of Yreka, and is optioned to O. Hauge.   It is in Sec. 24, T. 45 N., R. 8 W., four miles due west of Yreka, near the Mt. Vernon mine.   A quartz vein in greenstone averages 6 inches in width, with a maximum of a foot.   It strikes N. 20° W., and dips 70° W.   Development work consists of a 45-ft. shaft, the bottom of which connects with an adit level comprising 25 ft. of crosscutting and 100 ft. of drifting.   A second adit connects at the shaft at the 25-ft. level.   Drifting and crosscutting on this level total 85 feet.

Vein-quartz is kept separated from the country rock in mining, and it is stated to yield $25 per ton in gold in the mill.   Two stamps of 1000 lb. each are driven by an automobile engine.   Treatment is amalgamation only.   At the time of visit, a rented portable air compressor was in use.   Four men were working.

Bowser Mine is in Sec. 11, T. 47 N., R. 8 W., and is reached by mountain road from Hilt.   It is owned by the *Buschow Lumber Co.*, Hatfield, Arkansas, and is leased to C. L. Bowser of Hilt.   Two quartz veins, each about a foot in width, are separated by a basic dike, 18 inches in width. The strike is east and west, and the dip 60° N. The dike is highly altered, in part to serpentine. On both walls of this combination of veins and dike, the country rock is diorite. A crosscut adit, 167 ft. long, through diorite, reaches the vein, on which roughly 200 ft. of drifting has been done. Work at time of visit was in a small stope, 30 ft. above this level, from which both of the veins and the dike were being mined. Backs to the surface average 100 feet. Small lots of the quartz are treated in a Straub ball mill with a rated capacity of 10 tons per day. It is driven with a 6-hp. gasoline engine. Treatment is amalgamation on plates followed by concentration on a small Wilfley table.

Buzzard Hill Mine, Inc., owns the Buzzard Hill mine of 12 unpatented claims in Secs. 4, 5, T. 15 N., R. 7 E., also the Independence mine, which see. J. E. Merriam, Bedford Hills, New York, is president, and P. M. Tolman, Happy Camp, is superintendent. At the Buzzard Hill, hundreds of feet of development work have been done in gossan ore carrying gold and in the underlying body of heavy pyrite. In 1930, an attempt was made to extract the gold from the gossan with a 3½-ft. Huntington mill using plate-amalgamation. The mine was idle in 1933 and 1934. For additional details on development work see references given in accompanying table of mines. The nearest railroad point is Hornbrook, 80 miles east.

Cal Oro Dredging Co., see Gardella Dredge.

Central Mine in Sec. 2, T. 47 N., R. 8 W., and Sec. 35, T. 48 N., R. 8 W., is owned by C. R. Wiegel of Redding. The mine is reached by mountain road from Hilt. The property includes a 20-acre mining claim, 40 acres patented land adjoining the mining claim, and an 80-acre patented tract covering a spring for water supply.

A quartz vein associated with a basic dike is developed by three adit levels. Country rock is diorite. Dip of the vein is 66°. The No. 1, or highest adit level, has been stoped to the surface and has caved. No. 2 level, 65 ft. below No. 1, comprises 365 ft. of drifting. Wiegel states that there are three ore-shoots on this level, each 30 ft. long and from 3 to 6 ft. in width. No. 3 level is 75 ft. below No. 2, and is 367 ft. long. Wiegel is planning to install a 24-ton ball mill, flotation plant, and air compressor; and he was re-grading roads in order to haul in this machinery at time of visit in June, 1935.

Cherry Hill Mine, in Sec. 27, T. 45 N., R. 8 W., has been described in some detail in State Mineralogist's Report XXVII,* to which the reader is referred. Eleven claims are now held by G. A. Reichman of Fort Jones, and sons, Carl and Fred. Hundreds of feet of drifting and considerable stoping have been done on quartz veins one and two feet wide, with occasional swells to six feet. These workings, which are high on the hill, have been made more accessible with a new road since the last report was written. The 1800-ft. adit at creek level, or 450 ft. lower in elevation than any of the old workings, has been cleaned out. The property is well supplied with water and timber; and a power-line crosses it.

---

* Averill, C. V., *op. cit.*, p. 31.

China Point Placer Mine of 40 acres, patented, and 155 acres, unpatented, in Secs. 5, 6, 7, 8, T. 16 N., R. 8 E., near Happy Camp, is held by C. E. Reagan of Happy Camp. Three old drifts on the second channel above the present river, each 100 ft. or more long, are stated to have produced well. The unpatented portion of this property is for the purpose of covering a tunnel-site for a 1400-ft. tunnel, which will drain two miles of river channel. Low bars and river-bed are also covered by this location.

Classic Hill Mine (Placer) of 400 acres, of which 70 acres are patented, is held by C. Scott Greening of Happy Camp. It is in Sec. 36, T. 18 N., R. 6 E., near Indian Creek, 11 miles north of Happy Camp by a good road. About 60 acres of this property were mined off years ago by the hydraulic method with a yield of gold said to be very good. The material mined consisted of the decomposed, oxidized portions near the surface of a network of stringers and seams carrying quartz and free gold. The deposit is associated with a serpentine dike intruded into a series of slates, schist and limestone.

In 1934, an option had been given to Wason and Riley of Seattle, Wash., who were cleaning out the old ditch and repairing the flumes on the lower line, six miles long. Al Hahn was superintendent. Plans called for building a sawmill at the mine as soon as water-power from the lower ditch was available. Lumber was then to be sawed for building flumes on the upper ditch nine miles long. According to Greening, the lower ditch carries 1500 miner's inches, enough to operate three No. 2 giants with 4½-inch nozzles. Head is 140 ft. References to descriptions of earlier operations at this mine will be found in the accompanying table of mines.

Clyburn Placer is a property of 70 acres covering a gravel bar of the Klamath River in Sec. 8, T. 46 N., R. 7 W. It is owned by S. F. Clyburn, and is leased to C. W. Murphee of Kilgore, Texas. Early in 1935, considerable heavy equipment was installed here including a steam shovel, belt conveyor from shovel-pit to washing-machinery on the bank above, trommel 4 ft. in diameter by 15 ft. long, a stacker of linked buckets for coarse tailing, and a sluice for washing undersize from the trommel. Power for the washing-plant was furnished by gasoline engines, and water was pumped from the river with a semi-diesel engine. After the installation was completed, a few cubic yards of gravel were washed; but the plant was idle when visited in June, 1935.

Collins Ranch Placer (Ambrose Mineral Patent) consists of 89 acres in Secs. 6, 7, T. 16 N., R. 8 E., near Happy Camp, held by W. E. Collins, Seiad. Of this about 50 acres is placer gravel on a low terrace of the Klamath River, with a bedrock of dark, greenish-gray schist containing lenses of crushed quartz with widths of a few inches.

In 1933 and early 1934, Collins had mined about 750 cubic yards from a pit practically on the water's edge of Klamath River. The deposit is here 10 ft. deep including 5 ft. of sand overburden. The work was done with water pumped from the river by a 4-inch centrifugal pump and automobile engine. Some piping has been done with this water, but most of the gravel has been shoveled into the sluice box, which is one foot wide. Gold from this work is stated to have been sold for $400. No account was kept of prices received for different lots sold, but approximately half of it was sold at $16 per ounce and half at $26 per ounce. Gold is rough, and is usually in pieces worth 5¢ and 10¢. The largest pieces are worth $2 or $3.

An area of about two acres was worked on the south end of this property, 40 years ago, by Chinese. Depth here is 15 ft. This is a higher channel, 200 ft. east of the present river, and the bedrock is roughly 40 ft. higher than river level. A green dike (andesitic ?) in the schist forms the west rim of the pit. Water for this hydraulic mining was brought from Cade Creek.

Commodore Mine is in Secs. 26, 27, 34, 35, T. 46 N., R. 9 W., 30 miles by road from Yreka. It was described in State Mineralogist's Report XXVII, chapter for January, 1931, at which time eleven unpatented claims and a mill-site were optioned by H. J. Barton of Yreka to *Swedish American Mining Co.*

The mine is now being operated by *Goodenough Mining Co.*, Chas. DeForest, president, 1209 Smith Tower Building, Seattle, Washington. Harvey L. White of Walker (post office) is in charge, and he and three other men are at work, opening an old part of the mine known as the Davis workings. Old drift-adits of a length of 300 ft. have been cleaned out, and 100 ft. of new work has been done. A quartz vein carrying oxides of iron and free gold is exposed in widths varying from 12 inches to 22 inches. The dip is nearly vertical; and both walls are granodiorite. A basic dike, 75 ft. in width, carrying low-grade values in gold, occurs on the property. Present operators have cleaned out an old 250-adit that was driven to explore the dike.

In 1934, considerable work was done at this mine under the supervision of William L. Merritt, mining engineer, Railway Exchange Building, Portland, Oregon. Workings on the Commodore vein of a total length of 1000 ft. of crosscuts and drift-adits were cleaned out, and some new work was done. While these workings are open at the present time, they are not in use. In each of the two adits at the Insurance (see reference above), 100 ft. of new work was done. The basic dike was explored by means of 250 ft. of new tunnel driven entirely through a ridge.

Cornish Mine is on Big Humbug Creek northwest of Yreka. In 1930, a strike of good ore in a vein 28 inches wide was reported by F. M. McConochie and Henry Hayes. The mine has not been visited.

Cosmos Mines Development Co. is operating the Hansen mine, which see.

Davis Consolidated Mines (Placer) extend along Klamath River at Happy Camp for a distance of nearly three miles. They have been worked extensively by the hydraulic method during different periods from 1852 to 1919. The owner is Estate of Reeves Davis; address, W. F. Davis, executor, 427 J street, Sacramento.

According to W. F. Davis, the property now comprises 500 acres, patented, and 103 acres, unpatented, in Secs. 2, 10, 11, 15, 16, T. 16 N., R. 7 E., H. M. No land is now held in Secs. 21, 22, the ranch having been sold. He states that 400 acres of unmined ground remain, on the right bank of Klamath River only. Both sides of Indian Creek and both sides of Grider Creek are included. The ditch from Grider Creek was put in serviceable condition in the fall of 1932, and hydraulicking has been carried on during the seasons since. About one-quarter of an acre was mined off in the season, 1933–34, from a pit on Grider Creek, with a 20-ft. face. During the season, 1932–33, the location of the work was half a mile farther north on Curly Jack Creek, in a pit with a 30-ft. face. Three men were working.

The property is equipped with a sawmill, which will require reconditioning, and some other buildings.

In addition to this main property at Happy Camp, the Estate of Reeves Davis owns a 160-acre unpatented claim in Secs. 6, 7, T. 16 N., R. 8 E., a few miles up the river. This is primarily to cover a tunnel-site across the "boot bend" to drain the river channel for mining. However a prospect tunnel has exposed a high terrace containing gravel, 150 or 200 ft. above the present river. At a point some 40 ft. below, 300 or 400 cubic yards have been worked out with a giant as prospecting and assessment work.

Davis and Howard also hold 60 acres on the South Fork of Indian Creek, 14 miles above Happy Camp. This includes the *Jade* claim of 40 acres and an unnamed claim of 20 acres. On the latter is a 4-ft. vein of green *Californite*. Boulders of a good quality of this mineral are found in the creek. Two men are placer-mining for gold on the Jade claim. References to earlier reports on the Davis properties are given in the accompanying table of mines.

Dewey Mine is an old producer in Sec. 6, T. 41 N., R. 6 W., at an elevation of 6800 feet. The last of this old work was done in 1917. References to reports on it are to be found in the accompanying table of mines. Production during the earlier years of operation (about 1907) was from oxidized ores mined above the 100- and 200-adit levels. From Gazelle, the mine is reached by 10 miles of mountain road.

Dewey Gold Mines Trust, Gazelle, California, or 412 Chanin Building, New York City, has reopened the Dewey mine recently. The 300-adit level, at the same elevation as the top of the mill, has been cleaned out and retimbered where necessary, and roughly 2000 ft. of drifting and cross-cutting are now accessible. At a point 700 ft. from the portal is a winze on the vein, which goes down 150 ft. to the 450-level. Preparations are being made to unwater this.

The vein, consisting of quartz in widths as great as 12 ft. with walls of a basic diorite or a gabbro, strikes east and west, and apparently has a dip of 70° or 80° north but great variations in dip occur. Operators state that sulphide ore of good grade is known to exist in the bottom of the winze. Several hundred feet of diamond drilling have just been done to prospect in the walls of the main vein for parallel veins; and the results of this are stated to have been good.

The 10-stamp mill of 950-lb. stamps, with plates for amalgamation, two Gilpin County end-bump tables, and three Wilfley tables, has been repaired. The seven-mile power line has been rebuilt to provide electric power throughout. Capacity is 30 tons in 24 hours through a 24-mesh screen. This is to be used as a pilot-plant, and flotation will probably be added if laboratory tests are satisfactory. The mill building contains a single-stage duplex air compressor driven by a 50-hp. motor.

O. L. Scribner is mill superintendent. F. A. Wright, 806 California Building. Oakland, California, is manager.

Eliza Mine in Secs. 4, 5, 8, 9, T. 45 N., R. 8 W., is owned by R. H. DeWitt of Yreka. The mine is reached by means of a fair road, 15 miles long, which turns from the state highway at Hawkinsville, two miles north of Yreka. The last few miles are steep and narrow. A vein showing a width of from a foot to 10 ft. of quartz, averaging about 4 ft., is developed on the fifth level with 800 ft. of drifting. The strike is nearly north and south, and the dip is 70° west. At an elevation 224 ft. lower, more than 1000 ft. of work has been done in the country rock without picking up the vein. A raise connects the two levels; and short crosscuts from this should locate the vein, and point the way for additional development work. It is a typical fissure vein with ribbon quartz containing arsenopyrite, sphalerite and galena in small amounts.

O. H. Lawson, who was doing assessment work at time of visit, estimates the average value per ton at $7.60 in gold. The exposures are large enough to warrant systematic sampling and the preparation of an assay map, which is not available at present. Lawson states that the ore is all in place from the fifth level to the fourth, on which 500 ft. of drifting has been done, a vertical distance of 137 feet. All ore above the fourth has been stoped, except a length on the strike of 180 ft. where the ore was base. This ore was treated by amalgamation in an old 10-stamp mill, operated by water power, that stands on the property. Past production has been reported as about $150,000.

State Mineralogist's Report XXVII, chapter for January, 1931, contains a map of these workings on the two lower levels. References to earlier reports are contained in the accompanying table of mines.

Elliott Creek Mines Company (Placer) holds a property of 450 acres, formerly known as the *Daffodil*, on Elliott Creek just south of where it crosses the state line in Sec. 18, T. 48 N., R. 10 W., M. D. M., 40 miles from Medford, Oregon. The last six miles up Elliott Creek is trail. The property was idle when field work was done in this district, and it was not visited. Information given here was supplied by Harry Whitney, Copper, Oregon. W. L. Cobb, Roseburg, Oregon, is president of the company, and Dr. Sealy is secretary.

Gravels in the channel of Elliott Creek are covered by a big slide of schist. At the toe of this slide, on the opposite side of the channel, a point of solid schist juts out; and through this a tunnel, 565 ft. long, has been driven to reach the gravels. Washing is done in a sluice installed in the tunnel, with water brought through 2000 ft. of 3-ft. by 5-ft. flume.

Empire Group of 60 acres in Sec. 27, 34, T. 41 N., R. 9 W., M.D.M., 41 miles from Gazelle, is held by J. W. Wright of Etna. Claims are called Empire. May Day and Reed. There is a quartz fissure vein on each claim in a series of slates and cherts intruded by andesitic dikes. The May Day vein varies from 4 inches to 18 inches in width, averaging 8 inches. It strikes northerly and southerly, and has a vertical dip. A drift tunnel, 300 ft. long, gives backs of 50 feet. Ore has been removed from small stopes, 15 ft. long, and has been treated in an 8-ton Ellis mill on Sec. 27, owned by Dan Roff of Etna. It is equipped also with a crusher, and plates for amalgamation. The value of the ore is stated to be from $6 to $42 (gold at $20), free-milling, with some stringers giving $65. Sixty tons of ore that had been milled averaged $10 per ton. On the Empire claim, a similar vein has been developed with about the same amount of workings. Only a small amount of work has been done on the Reed, where arsenical ore is found.

In Sec. 27, a quarter of a mile to the north, Wright holds the Lena claim. A 75-ft. dike, probably andesitic, contains quartz stringers showing plentiful chalcopyrite. Copper carbonate weathered from this has worked down through the fractures, and has stained the dike for quite a width. A tunnel started on the slope of a hill near the middle of the dike shows this stained dike material for 50 ft., then passes into the slates and cherts. Wright states that this 50-ft. width of dike assays $3.20 in gold (gold at $20). Five stringers in the dike, width 3 or 4 inches with a maximum of 18 inches, are said to average $25 per ton in gold. In a surface cut, a sample including a 4-inch stringer and silicified dike material to make a total width of 5 ft. is stated to have assayed $7 per ton in gold. A separate sample, 8 or 10 inches wide, including the stringer, gave $15.60 in gold, according to Wright.

Enterprise Placer of 100 acres, in Sec. 21, T. 45 N., R. 10 W., on the edge of the town of Scott Bar, is held by Hollis Anderson of Scott Bar and others. The property covers portions of four separate terraces of Scott River. Recent work includes a 50-ft. prospect shaft, of which the collar is 300 ft. higher in elevation than the present river; also a 26-ft. shaft at an elevation 60 ft. still higher. The material excavated carries numerous fragments of Abrams mica-schist; and these are said to be associated with the occurrence of placer gold. The bedrock is also of this schist, and contains small veins of quartz similar to the larger ones at the Quartz Hill mine nearby. These small veins also carry gold.

In 1855, an 1100-ft. adit was driven from river-level, and it was reopened in 1888. Some ground carrying high values in placer gold is stated to have been found; but caves and bad air drove the miners out. See State Mineralogist's Report VIII, p. 623.

Fledderman Bros. of Yreka are re-working tailing from old drift mining at the mouth of Greenhorn Creek, behind the Greenhorn school just south of Yreka, in Sec. 34, T. 45 N., R. 7 W. The drifting was done 35 years ago at a depth of 130 ft. below the surface. The gravel contained considerable clay, and it was not thoroughly washed in the original operation. The clay has now disintegrated to such an extent that the gravel washes well. Operators state that a recovery of 14¢ per cubic yard is being made, and that this is profitable. Digging is done with a gasoline power shovel of the dipper-stick type, mounted on caterpillar treads. The dipper has a capacity of a third of a cubic yard. All of the washing-machinery is mounted on a truck, and is driven by the truck engine. There is a hopper to receive gravel from the shovel, then a feeder driven by an eccentric to deliver a steady stream of gravel to the revolving trommel. This is punched with ½- and ¾-inch round holes. Fines are washed in a sluice, and oversize is stacked by a conveyor belt. With three men working, 40 cubic yards are handled per hour; but only two men were workng at time of visit. A total of 16 gallons of gasoline is burned per 8-hour shift by the two machines. Water is brought in by gravity by means of ditch, flume and pipe from Greenhorn Creek. Prior agricultural rights to this water make the working-season short.

Fledderman Bros. washing tailing from old drift mining near Yreka.

Gardella Dredge (Cal Oro Dredging Co.) near Yreka.

Gardella Dredge. *(Cal Oro Dredging Co.)* A dredge with 5-cu. ft. buckets has been operated by Lawrence Gardella just south of the town of Yreka, and on the east side of the Pacific Highway during 1933, 1934 and 1935.

Gilta Quartz Mine is an old producer in Sec. 12, T. 9 N., R. 7 E., eight miles by trail south of Forks of Salmon. It is owned by G. A. Dannenbrink and others of Etna. In 1931, James Tracy of Forks of Salmon and Paul A. Bundy, Mining Engineer, were conducting a three-month prospecting campaign here with a crew of four miners. References to earlier reports on operations here will be found in the accompanying table of mines.

Gold Ball Mining Company is operating on the Stevens Group in T. 39 N., R. 11 W., at the head of Eddys Gulch near Sawyers Bar. F. A. Gowing, c/o Jas. M. Allen, Yreka, is president and general manager of the company, and A. F. Chapman is secretary. The mine is reached by 67 miles of road from Montague or 61 miles from Yreka. There is paved road for 31 miles, Yreka to Etna, and the balance is mountain road with grades of 15% and 20%. A 6000-ft. summit is crossed west of Etna, and when this is blocked by winter snows, a longer route by way of Somes Bar is used. Names of claims are Specimen, Blue Bell, Banner and Buster. Apparently part of it is a relocation of an old claim called the *Wisconsin*, which has been mentioned in earlier reports in connection with the *Mountain Laurel*.

Ore is being mined from a flat vein showing widths of 4 ft. to 5 ft. of quartz. It is developed by an intermediate level at the top of a 60-ft. raise from the Stevens adit. Treatment is in a new 25-ton mill containing 10-inch by 12-inch crusher, 4-ft. by 4-ft. Marcy ballmill, Dorr classifier, plates for amalgamation, and two Kraut flotation cells. Power is furnished by 100-hp. diesel engine, which drives a 260-cu. ft. compressor in addition to the mill. Considerable graphite from the graphitic schist country rock goes into the flotation concentrate. Mining has been done in this immediate vicinity at different times under the same name as that given above or similar names, and references to earlier reports will be found in the accompanying table of mines.

Golden Eagle Mine in Sec. 11, T. 44 N., R. 9 W., 10 miles north of Fort Jones, is owned by George A. Milne of Fort Jones, who owns several other patented mining claims and other patented ground adjoining. The mine has been worked at various times for many years, and total production is estimated at $1,000,000 by the owner. In 1927, the *Sterling Gold Mining Co.*, backed by Oakland people, did some work here; but little production resulted. The mine is developed with a steeply inclined shaft, the lowest level of which is 225 ft. below the surface. On this level more than 2000 ft. of work has been done, mostly drifting on veins with an average width of 2 ft. Milne states that the best ore came from a stope 200 ft. south of the shaft, now caved. The shoot is said to rake to the south; and Milne thinks that the best way to open up new ore is to sink the shaft and open up this shoot below. A map showing this lower level and its geology is contained in State Mineralogist's Report XXVII, chapter for January, 1931. Reference to earlier reports are contained in the accompanying table of mines. The only recent work here has been the milling of ore from the nearby New York mine.

The old mill from the New York mine has been moved to the Golden Eagle and set up with another mill. The combined plant now consists of the following: rock crusher, six 1000-lb. stamps, ten 850-lb. stamps, amalgamating plates, and an Overstrom table. This machinery is driven by electric motors of 40, 50 and 5 horsepower. At the mine are a 50-hp. electric hoist, Giant compressor, 2 cylinders, single stage, driven by a 50-hp. motor, three jackhammers, ore cars and tools. A 2-inch, 4-stage centrifugal pump is used to keep the shaft unwatered; and three air-driven pumps are on hand. A small assay office, fully equipped, also stands on the property.

Gold Leaf Mine of 80 acres in Sec. 26, T. 46 N., R. 7 W., six miles north of Yreka, is owned by J. M. Weaver of Yreka. It is developed by a 185-ft. tunnel and a 50-ft. winze. According to the owner, some gold ore of very high grade has been produced recently from quartz 2 inches to 12 inches wide in a 20-inch vein. When field work was done, the mine was idle and the winze was full of water, so it was not visited.

Good Hope Claim adjoins the Gilta to the south, and is held by Sam R. Israel, Forks of Salmon. It has not been visited.

Gravel Mines, Ltd., is operating a placer property known as the *Frank* (Shorty) *Lewis* mine in Secs. 5, 8, 9, T. 16 N., R. 8 E., H. M.,

Gravel Mines, Ltd. Hydraulicking on high terrace of Klamath River near Happy Camp.

about six miles east of Happy Camp. Five unpatented claims of 20 acres each are held. C. C. Broadwater, 27 Highland Ave., Piedmont, is president, and H. S. Shuey is secretary of the company. J. C. Moulton, Seiad, is in charge at the mine.

Gravel to be mined is on a terrace of Klamath River, 110 ft. higher than the present stream. Water is brought in from China Creek through 4200 ft. of 36-inch by 32-inch flume and 5400 ft. of ditch. Capacity is 1000 miner's inches, which is supplied to one No. 3 giant under a head of 175 ft. At time of visit, a 30-ft. depth of gravel was exposed in the pit recently opened. Bedrock is a slaty schist. Gravel is washed in 100 ft. of sluice, 26 inches by 26 inches, with block riffles, and plans call for the installation of some riffles of angle-iron; also undercurrents.

Gumboot Mine in Sec. 16, T. 45 N., R. 9 W., consists of five unpatented claims held by Jack McInnes and Philip McCool, both of Scott Bar. It is seven miles by road east of Scott Bar; but the last 1½ miles are too steep for an automobile and are used as sled-road only. Upper workings consist of a crosscut adit, 120 ft. long, passing through a series of impure limestones and slates, striking S. 22° E., and dipping 60° southwest. At the 120-ft. point is a drift on the vein 100 ft. in length. A raise from this drift connects with the surface, length 100 ft. measured on the dip of the vein, which strikes N. 55° W., and dips from 32° to 62° to the southwest. It varies in width from nothing to 5½ ft., with an average of 3 ft. A shipment of 18 tons from half way up this raise, said to have been unsorted, is stated by the owners to have brought gross returns of $75.73 per ton, 6 oz. being in silver, the balance gold (@ $20.67 per oz.). The ore is white quartz showing oxides and galena. The hangingwall is rhyolitic; and the footwall is calcareous slate. A sketch-map showing these workings is contained in State Mineralogist's Report XXVII, chapter for January, 1931.

Other veins show in shallow prospect cuts on the surface. One of these strikes S. 40° E. and dips 45° southwest. At an elevation 70 ft. below, 300 ft. of crosscut-adit has been driven through calcareous slate in an attempt to cut this vein on its dip. At an elevation 200 ft. below the workings first mentioned above, is a 200 ft. tunnel, which was started about 35 years ago to tap the veins. It passes through a metamorphosed sedimentary series, quartzite, limestone and slate.

Guy Ford Group of Mines—See Jumbo.

Hansen Mine in Sec. 1, 12, T. 9 N., R. 7 E., H. M., seven and a half miles by trail south of Forks of Salmon was being operated, late in 1934, by *Cosmos Mines Development Co.* The crew consisted of 10 men under Ralph Bender. A mill, said to be of 30 tons daily capacity, had recently been installed. The property was not visited. References to earlier reports are given in the accompanying table of mines.

Hathaway Mine. The following is quoted from State Mineralogist's Report XXVII, chapter for January, 1931:

"It comprises the NE¼ Sec. 11, T. 40 N., R. 9 W., which is assessed to Albert S. Hathaway, *Sugar Creek Mining Co.*, 410 Holm Boulevard, Los Angeles. The *Sugar Creek Mining and Milling Co.*, C. Ray LaMar, president, 365 S. Cloverdale, Los Angeles, has done a little recent work on the property. At the time of visit, M. P. Mullen, Gazelle, California, was the only one there. A 340-ft. tunnel passes through impure limestone; and a drift has been driven on a vertical seam for a distance of 50 feet. The seam strikes N. 20° E., and has a width of one foot. The filling is a highly oxidized gouge, showing green copper staining and specks of sulphide. Mullen stated that this assays $10 to $12 per ton, and that, years ago, it produced $60,000 from a stope 95 ft. high, 85 ft. long and 30 inches wide. A raise from this lower level taps the stope; and 94 ft. above a second tunnel connects with the stope. The seam is in a basic dike, of which only one wall is exposed. The part of the dike that can be seen in the workings measures 32 ft. in width; and it has the same strike and dip as the seam mentioned above. The series of limestones, slates and cherts strikes S. 74° E., and dips 61° S.

The Stevens tunnel is 400 ft. long; and it exposes a dike of similar basic rock, showing calcite seams, iron-oxide stain, and green copper stain. Mullen stated that an 82-ft. width of this assayed $4 per ton in gold. On railroad land in the SW¼ of Sec. 11, on which the company was said to have an option, is a vein with a maximum width of 5 ft., on which a tunnel 420 ft. long has been driven. The strike is N. 26° E., and the dip 57° to 80° northwest. The vein is similar in appearance to that first described; but the dark colored dike is lacking. A high content of lime also tends to lighten the color. The vein matter is highly stained with brown oxides of iron, and in places shows a high sulphide content. An 85-ft. raise was said by Mullen to have given an average assay return of $31 per ton for a 30-inch width. The footwall is a gray chert; and the hanging wall is an impure limestone.

A 45-ton ball mill and concentrating tables were on the property, also a two-drill compressor, 10 by 12 inches, with two receivers. None of this machinery had been installed."

Since the above was written, some work has been done here by *Northern California Gold, Inc.* (see under Morrison & Carlock). However, when field work was being done in 1935, the property was idle; and it was not visited.

Hazel Mine, formerly *Jillson,* in Sec. 31, T. 47 N., R. 6 W., and Sec. 25, T. 47 N., R. 7 W., four miles southwest of Hornbrook, is the property of the Hazel Gold Mining Co., C. F. Kimball, secretary, 1103 First National Bank Building, San Francisco. Quartz veins, 3 ft. wide, carrying gold are found near the contact of the metamorphic series, called in this report Paleozoic and pre-Paleozoic sediments and schists, and the greenstone. Greenstone is the more prominent formation on the surface; but the dumps indicate that much of the work may have been in other formations. The mine has been idle for some time; and the workings were not examined. Reports of the State Mineralogist (see table of gold mines) state that the walls are slate. With the exception of one or two old buildings, and a 14-in. by 9-in. by 10-in. compressor, two-stage, connected by belt-drive to an 85-hp. motor equipped with starter, little remains on the property. The mill has burned. An interesting exposure of the basal conglomerate of the Chico formation is seen a quarter of a mile east of the mine. It is composed largely of pebbles and cobbles of various igneous rocks, andesites, diorites, greenstone, etc., also some chert. To the west of the mine workings is an outcrop of white and gray chert.

Herndon Placer Mine, in Sec. 17, T. 46 N., R. 7 W., is owned by A. C. Herndon of Hornbrook. A description of a former operation here is contained in State Mineralogist's Report XXVII,[*] to which the reader is referred. In 1935, a test of this property was being made by C. S. Lowe with a slackline excavator, using a steam donkey engine for hoisting. The mast supporting the cables of the excavator stands just

Herndon Placer. C. S. Lowe operating slack-line excavator and washing-plant.

behind a hopper, into which gravel is dumped. It is washed on a grizzly of 35-lb. rails with 2-inch spacing, and undersize passes to a sluice with riffles. Water is pumped from Klamath River at the rate of 700 gallons per minute with centrifugal pumps driven by two model-A Ford gasoline engines. With six men working, 150 cubic yards of gravel are washed per day.

* Averill, C. V., *op. cit.,* p. 58.

A year or two ago, an attempt was made by *Cosul Company* (Chinese) to work this property by the old method of developing power with current-wheels in the river. One wheel was used for hoisting gravel from the pit in a wooden skip. A second wheel drove a Chinese pump to remove water from the pit. Apparently the capacity of this outfit for moving gravel was extremely limited.

Photo by Walter W. Bradley.

Current wheels of Cosul Company (Chinese) in Klamath River, on Herndon Placer, 1933,

Hickey Mine, eight miles southeast of Sawyers Bar, is a property of six mining claims held by Edward Hickey of Sawyers Bar and San Francisco. It has been worked in a small way for many years, and a reference to a report on earlier work is to be found in the accompanying table of mines. For the past seven years, it has been leased by Jack Blaylock. He hunts out lenses of high-grade quartz in the decomposed formation, and treats the ore in a 4-ft. arrastre operated by hand power. The last 75 tons are said to have yielded a return of $35 per ton.

The deposit is apparently a network of stringers of quartz, an inch or so wide, in the country rock, probably a schist. The country rock has decomposed to clay, so that the deposit now is a loose, disintegrated mass of fragments of quartz mixed with clay. The depth to which this condition extends is unknown, but a tunnel has indicated a depth of 100 feet and more, in one place at least.

Lee DuBois of Yreka, who holds an option on the property, believes that this deposit contains enough gold to be profitably worked by power shovel for a length of 3800 ft. and a width of from 40 to 85 ft. He has made a preliminary sampling with this end in view. He states that 70 samples taken to a maximum depth of 6 ft. with a post-hole auger gave an average return of $3.34 per ton with gold figured at $20.67 per ounce. As much as four feet of soil overburden was found in places. A 2500-word report on this property, by DuBois, is on file at the Redding office of this division. There is a similar occurrence of gold on the Keaton property adjoining. A new road has been built to the Jumbo and Lanky Bob mines, within two miles from the Hickey and Keaton.

Hi-You Mine of three claims, the Hi-You, Aurora and Apex in Sec. 30, T. 45 N., R. 8 W., is held by G. A. Reichman and W. H. Young of Fort Jones. There is a good dirt road to Hi-You gulch, but the last mile to the mine is steep sled road. The only tunnel now open exposes a flat vein of ribbon quartz with a dip of 25°. In length this adit totals about 200 feet. The white quartz occurs in bands from half an inch to an inch wide, separated by very thin seams of graphitic material. Width of the vein is 1 ft. to 2 ft., averaging where seen by the writer about a foot. It outcrops with a width greater than 2 ft. at a point on the ridge about half a mile from the main workings. The quartz carries free gold and pyrite. Walls are graphitic schist, and considerable pyrite is seen on some of the fractures in this country rock. On the hanging wall, a very fine-grained dike, probably andesitic, is found for a short distance. Five other adits, now caved, have been driven. They vary in length from 70 ft. to 200 ft.

Ore from these workings has been milled at the 10-stamp mill on the Hi-You placer claim, a mile below, by the sled road. The stamps weigh 1000 lb. each. Treatment is amalgamation followed by concentration on a Wilfley table. A rock crusher and a 12-hp. gasoline engine complete the equipment. Returns from the ore have varied from $2 to $20 per ton.

Hi-You Placer on the gulch of the same name, and just below the quartz mine, is also held by G. A. Reichman and W. H. Young of Fort Jones. In May, 1935, it was being equipped with a large portable washing plant, including a trommel, to wash gravel dug by a power shovel. This installation was being made by a lessee.

Huey Mine is in Sec. 25, T. 18 N., R. 6 E., H. M., on Indian Creek above the Classic Hill mine, and about a mile distant. The deposits at the two mines are similar. David Huey and one other man work with a No. 2 giant supplied with water from Mill Creek through a three-mile ditch. References to this mine are given in the accompanying table of mines.

Hunter's Paradise—see Paradise.

Independence Mine in Sec. 32, T. 15 N., R. 7 E., H. M., 14 miles by road south of Happy Camp, is owned by *Buzzard Hill Mine, Inc.* J. E. Merriam, Bedford Hills, New York, is president of the company, and P. M. Tolman, Happy Camp, is secretary. Seven quartz locations are held on Independence Creek at an elevation of 1225 ft. Hornbrook, the nearest railroad station, is 87 miles east.

The mine is a pocket-mine, which produces gold from high-grade pockets in a quartz vein striking N. 45° W., and dipping about 60° southwest. Widths average 2 ft. to 3 ft., with maximums of 6 ft. to 8 ft.

Total production to date has been roughly $300,000. Country rock is a basic schist intruded by andesitic dikes.

Workings and equipment of this mine were described in State Mineralogist's Report XXI, p. 446. Since that report was written, the shaft has been deepened to the 180-ft. point, and there are four levels from it with a total of about 750 ft. of drifting. Drifting on the lowest level amounts to 40 ft. in one direction and 30 ft. in the other. A road, 4000 ft. long, from Klamath River to the mine has been completed; also 6000 ft. of flume with capacity of 500 cu. ft. of water per minute, giving 220 ft. of head at the mine. Water-power is used to operate the 10-stamp mill with stamps of 1250 lb. each, and to drive a 250-cu. ft. Ingersoll-Rand Compressor. There is also an old 450-cu. ft. compressor in poor condition. Buildings include a good bunkhouse, boarding house, and manager's house. Mill-runs have indicated that it does not pay to mill the quartz between pockets. When checked in 1934, this mine had last been operated in 1930 and 1931.

Indian Girl Mine of 40 acres, in Sec. 14, T. 46 N., R. 7 W., is leased to Jack Clute, Nick Kuntzler, Wayne Whitney and Hoyt McClain of Klamath River. It is just north of the Klamath River highway, but several hundred feet higher in elevation. The four lessees are hunting for pockets on the property, which has the reputation of having yielded numerous good pockets in the past. On the slope of the mountain is quite an accumulation of dump material from cuts and tunnels made by pocket hunters. Part of the work of the lessees consists of screening the surface detritus, including these dumps, and of washing the fines in a long tom. Water is accumulated behind a dam in one of the old tunnels. They state that an average return of $3 per cubic yard in gold, is obtained from the surface material.

Iron Dike Mine of four claims on North Hungry Creek, in Sec. 22, T. 48 N., R. 8 W., is held by D. M. Watt and brother of Phoenix, Oregon. The mine is reached by mountain road from Hilt, or by the road leaving the Klamath River at Beaver Creek. A 175-ft. crosscut in diorite reaches a vertical quartz vein 2 ft. in width, on which there is 40 ft. of drifting and a small stope. The quartz carries gold in occasional high-grade pockets, and some of the basic dike rock associated with the vein is said to carry gold. Quartz is hauled in a truck to a $3\frac{1}{2}$-ft. Huntington mill about a mile down the mountain. Rated capacity is 10 tons per day. Three men are at work.

Jim's Seattle Placer Mine is a patented claim of 20 acres in Sec. 5, T. 16 N. R. 8 E., about five miles east of Happy Camp, owned by J. A. King, Happy Camp. It adjoins the patented 40 acres of the China Point Placer. Some production has been made here from the second channel of the river, about 50 ft. above the present stream. Prospecting for the next higher, or third channel, is in progress. The work consists of a 20-ft. shaft, all in overburden.

Johnson and Lewis group is a group on which some work has been done by the *Superior Consolidated Mines Co.*, 920 Lloyd Building, Seattle, Washington, with the late S. W. Steffner of Fort Jones as superintendent. The Johnson includes two patented claims assessed to

John J. Johnson and others of Etna; and the Lewis comprises four unpatented claims adjoining. The location is Sec. 18, 19, T. 43 N., R. 9 W., near the Morrison and Carlock mine. A quartz vein, 2½ ft. wide, is exposed on the surface for 1000 ft., exposures not being continuous, but at intervals only. Free gold can be seen in some of this quartz. A drift-tunnel on the vein, 110 ft. long, gives a depth of 110 ft. below the surface at the face, and, if continued to 1000 ft. will give 250 ft. of backs. An 80-ft. crosscut tunnel at about the same level has not yet reached the vein. Relations between the various kinds of country rock are complex here, slate, limestone, andesite and rhyolitic breccia being noted. The company named above has also controlled the Morrison and Carlock and the New York mines (which see).

Jumbo Mine is a property of eight claims in the northeast part of T. 39 N., R. 11 W., six miles by road southeast of Sawyers Bar. The part of the road up Whites Gulch to this property and to the Lanky Bob, half a mile beyond, has just recently been built. The Jumbo is owned by Guy Ford of Weed, and is optioned to Lee DuBois of Yreka. It has been worked at different times in the past, and references to reports on these earlier operations are to be found in the accompanying table of mines.

An old ten-stamp mill of 1050-lb. stamps, which still stands on the property, was supplied with ore from workings on the west side of Whites Gulch. The portal of the main adit level has caved, but the vein can be seen at the surface. The quartz is of the pulverulent, or highly fractured variety known as 'sugar quartz'. Water power for the mill was furnished by a mile of ditch that delivered water at a head of 286 feet. DuBois has recently been repairing this.

On the east side of the gulch are a number of old tunnels varying in length from 70 ft. to 400 ft. They expose at least two veins, one a flat vein with an average dip of less than 30°, the other nearly vertical. On the latter is an 86-ft. adit, from which a 44-ft. winze has been sunk. DuBois says that samples taken in the vicinity of this winze, with widths of from 3 to 5 ft., average $20 per ton with gold figured at $20.67 per ounce. He drove 68 ft. of crosscut to reach this vein at a point 150 ft. lower in elevation, but estimates that it must be driven an additional 100 ft. Country rock in this vicinity is chloritic schist, apparently developed from an old series of andesitic lava flows. A 1500-word report by DuBois on this property is on file at the Redding office of this division.

Kanaka Hill Hydraulic Mine of 160 acres of unpatented mining claims in Sec. 27, 33, T. 16 N., R. 7 E., H.M., four miles below Happy Camp, is held by Steve S. Green, Happy Camp. It covers gravel of a high terrace of Klamath River, some 250 ft. higher than the present stream. Water supply is taken from Kanaka Gulch, which flows during storms and for a few days following storms. A No. 2 giant is supplied through 400 ft. of 11-inch pipe. This is sufficient for small-scale operations only. A better water supply can be obtained from Wilson Creek, one and a quarter miles to the south. At time of visit, pits roughly 500 ft. long, 30 ft. wide, and showing a 14-ft. face of gravel, had been opened.

Katie May Mine is in Sec. 13, 24, T. 45 N., R. 8 W., in the Greenhorn District, near Yreka. Andy Calkins of Horsecreek (post office) holds six claims, of which four have been held for many years. Bob Macaulay and Jack Perdue are leasing. They state that years ago some $70,000 was produced from a stope 40 ft. long on a 10-inch vein; and this is confirmed in a general way in the references gven in the accompanying table of mines. The lessees have recently made a new discovery on an entirely different part of the property. They have made cuts totaling roughly 50 ft. in length, and 4 ft. deep, extending both above and below a CCC road that has recently been built through the property. Pieces of quartz, an inch in diameter, contain visible free gold in considerable quantity. At time of visit, it was difficult to tell the width of the deposit, because the bottom of the trench was still in loose material. A width of 18 inches of this is said to give good pannings.

At a distance of 800 ft. from this strike, on the opposite side of the high ridge, is an 85-ft. tunnel on a 9-inch quartz vein. Macaulay states that 9 tons of sorted ore from this yielded $248 in free gold only. It contains auriferous chalcopyrite, which was not saved by the mill on another property, where it was treated. At another locaton on the property, Macaulay sank a 23-ft. shaft on a 3-ft. quartz stringer zone, which is stated to have yielded $3,100 in gold. Country rock is schist, of which a part has the appearance of slate.

Keaton Mine is a property of five claims in T. 39 N., R. 11 W., nine miles southeast of Sawyers Bar, owned by James Keaton of Sawyers Bar and optioned to Lee DuBois of Yreka. The occurrence of gold here is similar to that at the adjoining Hickey mine (which see), and the Keaton has also been worked on a small scale, with short tunnels and open cuts, for many years.

In 1927, an Ellis mill with a rated capacity of 10 tons in 24 hours, Blake crusher, and plates for amalgamation were installed; and runs from two open pits were made. According to DuBois, 100 tons from a pit just above the mill yielded $5.20 per ton, and 65 tons from an upper pit yielded $3.24 per ton (gold at $20). The last two miles to this property must be traveled over a steep sled-road, and this was used in the transportation of gasoline to drive the 7-hp. engine. Hence the operation was not profitable.

DuBois believes that a deposit large enough to mine with a power shovel can be developed here, with a length of 1600 ft., width 40 ft. to 120 ft. and depth 50 ft. or more. With this end in view, he took 108 samples with a post-hole auger to a maximum depth of 6 feet. These are stated to have given an average return of $3.60 per ton with gold figured at $20.67 per ounce. In one place a width of 120 ft. was sampled with a row of 10 holes. A 2000-word report on this property, by DuBois, is on file at the Redding office of this division.

King Jade Quartz Mine of six 20-acre claims in Sec. 7, T. 17 N., R. 7 E., H. M., is held by Harry D. Maltis of Happy Camp. The owner states that recent prospecting has disclosed a 6 to 8-ft. quartz vein that pans well in gold. Considerable *californite* ("jade") has been found also. The property was not visited.

King Solomon Mines Company is operating in Sec. 14, T. 38 N., R. 12 W., 8 miles south of the Black Bear post office. Roy N. Bishop, 411 Crocker Building, San Francisco, is president of the company, and Harry M. Thompson, address, Black Bear, is in charge of operations. W. C. Baldwin is mine foreman, and Wm. Samuels is mill foreman.

Photo by Walter W. Bradley.

King Solomon Mines Co., "mining" with caterpillar tractor and bulldozer scraper.

The mine is about 75 miles from Yreka, the first 31 miles, Yreka to Etna being hard-surfaced road, the balance mountain road with grades of 15% to 20%. The last eigth miles from Black Bear to the mine is practically all steep grades and sharp turns. This stretch was built by the present operators of the mine. When the 6000-ft. summit west of Etna is closed by snow, a route by way of Somes Bar is available.

Holdings of the present company are approximately 800 acres, including the old King Solomon holdings. Five of the claims are patented. The old work was done in the years 1900 to 1905, when four two-stamp batteries and a 5-ft. Huntington mill were in use. These 1000-lb. stamps, which were driven by steam power, are still on the property, but are in poor condition. Treatment was by amalgamation only. In this plant, 80,000 tons of surface ore were treated from a pit 560 by 100 ft. by 40 ft. deep. Recovery from this ore is said to have been $300,000.

Present operations were started in 1931, with an extensive campaign of prospecting and sampling by a method somewhat out of the ordinary.

An extension of the deposit formerly mined was developed by means of cutting trenches with road-building machinery, road-plow and 30-hp. Caterpillar tractor with scraper blade mounted in front (bulldozer). In this way 2000 ft. of trenches, 9 ft. wide, and of an average depth of 10 ft., were cut. Sampling by two independent methods was carried on at the same time. First, samples were taken by an expert panner, who made a rough estimate of the gold content by panning. The second method consisted of cutting 250 lb. of sample from each $2\frac{1}{2}$-ft. width of ledge matter exposed in a trench. This 250-lb. sample was taken to the assay office, and crushed to three-quarter-inch size, then reduced to 40-lb. with a Jones-type sample splitter. The 40-lb. sample was crushed to one-quarter-inch size and again split, then 10 lb. was ground to 40-mesh in a disc-pulverizer. A 4-lb. split was then charged to a 12-inch Geo. White muller for amalgamation with a drop of mercury. After grinding for five minutes, tailing was panned off and saved for a composite sample from each trench. The amalgam was parted with nitric acid, and the gold was weighed on an assay balance. A 4-lb. sample from the composite tailing sample from each trench was subjected to a second

Photo by Walter W. Bradley.

Mill and aerial tramway terminal, King Solomon mine.

grinding of 10 minutes in the muller, and considerable additional extraction obtained. A fire assay was made of tailing from this second amalgamation test.

Mining of the deposit so developed is now in progress. The earlier mining was done partly with a 50-hp. Caterpillar tractor with scraper blade (bulldozer), particularly the stripping of the first 20 ft. of depth. This machine pushed the surface material over a bank, so that a power shovel could be used to load it on trucks. Present mining is done entirely with a ¾-yard shovel powered with a diesel engine. Trucks deliver the ore to the head of a 2500-ft. aerial cable tram, where the primary crushing is done. In this way, five men are able to mine 300 tons in 8 hours, including two men to run the cable tram. On the latter are 16 buckets of one-half cubic yard capacity each, delivering ore to the mill.

The new mill contains two Hardinge pebble mills in closed circuit with a 20-mesh vibrating screen. Due to the softness of the ore, 70% of the output of these mills will pass a 200-mesh screen. Pulp is treated on large amalgamation plates, then goes to Kraut flotation machines. On account of the varying character of this ore, due to differences in oxidation and weathering in different parts of the deposit, operators have experienced difficulty in keeping the flotation process adjusted. They are now considering a cyanide treatment for amalgamation tailing. Power is furnished by an engine rated at 250 hp., which delivers about 200 electrical horsepower; and motors are used to drive the various machines. The engine burns 250 gallons of 27° Bé. oil per day. Storage of 43,000 gal. of this diesel fuel is provided for winter.

The ore occurs in a schist, which was probably originally a sedimentary rock. It has been intruded by dikes, probably dioritic, in such a way that, as one crosses the lode, a change in formation is noted every few feet. Quartz, calcite, and actinolite? occur in seams and vugs. The ore is all highly oxidized and shows the red, brown and yellow colors of oxides of iron, also some black oxide of manganese. According to the operators, the gold is all in the schist and the dikes are barren. In addition to these formations, which were seen in the orebody being mined, several others were noted in the immediate vicinity. On the ridge behind the mine is a basic igneous rock, possibly gabbro, intruded by an occasional fine-grained, nearly white dike, probably rhyolitic. One dike contains feldspar phenocrysts about a quarter of an inch long, and is apparently andesitic. A little limestone float was noted. The basic rock (gabbro?) is stated to have produced some good pocket-gold.

See the accompanying table of mines for references to notes on earlier work at this property.

King Tut Mine of three claims, in Sec. 32 T. 41 N., R. 7 W., is held by T. A. Lloyd of Gazelle. He holds an option on the adjoining SE. ¼ Sec. 29, also. An old shaft, said to be 55 ft. deep has been cleaned out by Lloyd to the 28-ft. point. It follows a nearly vertical fracture zone in a series of andesites and diorites, possibly gabbro. Occasional bunches of quartz show specks of chalcopyrite and considerable green copper stain. Lloyd states that he has obtained assays of $5.95 and $11.20 in gold from pieces of this quartz.

Klamath Placer Mining Co. is mining the *McConnell Bar* on Klamath River, six miles west of the Pacific highway. The property comprises three patented mining claims of a total area of 68 acres in Sec. 16, T. 46 N., R. 7 W. Officials include W. D. Miller, president, Klamath Falls, Oregon; L. J. Troy, secretary, Santa Monica, California; C. T. Phillips, vice president, in charge of washing operations; and Lee L. Parker, foreman.

Mining is being done in a pit 14 feet below river level with a one-yard gasoline shovel. A strip of gravel is left between the pit and the river, and seepage water is kept pumped out. The pit is an extension of one started in 1910, and it is now 1700 ft. long. Gravel is loaded by the power shovel into two three-yard dump trucks, hauled a maximum of 1000 yards, and dumped into a three-yard skip running on inclined 16-lb. rails. These are mounted on a trestle, 370 ft. long at a grade of 24% to reach the top of the washing plant outside of the pit. At the top is a gasoline hoist, which raises the skip and dumps the gravel onto a grizzley with eight-inch openings. Capacity is 500 cubic yards in eight hours, handled by nine men.

Gravel that passes through the grizzley is washed in a sluice containing 147 ft. of block riffles, set at a grade of eight inches in 12 feet. Seepage water from the pit is delivered to the top of the washing plant

Klamath Placer Mining Co.   Tram from pit, and washing-plant.

by a 5-inch centrifugal pump working against a head of 60 ft., and driven by a 25-hp. engine. This burns 2½ gal. of stove oil per hour, costing 10¢ per gallon. Additional water for washing is taken from the river by a 10-inch centrifugal pump working against a head of 50 feet. It is driven by a 52-hp. diesel engine, which burns 4 gal. of fuel per hour, costing 8¢ per gallon. Sands below $\frac{3}{16}$ inch in size are passed through the bottom of the main sluice by a grizzley with openings of that size, for treatment on an undercurrent. This is a table, 8 ft. by 12 ft., set on a slope of 15 inches in 12 ft., and equipped partly with Hungarian riffles and partly with angle-iron riffles. Some recovery is made in sluices below the undercurrent on ¼-inch screen over burlap. Costs for the entire operation are stated by the operators to be 22¢ per cubic yard including all depreciation and overhead.

They have made assays and a mill-test of sand-tailing from the undercurrent, and state that it contains $4 per ton in gold. A pebble mill is to be installed to grind the sand for a further recovery. Plans also call for later turning the river through the pit to drain the present river channel for mining.

Knownothing Placer is a tract, in Sec. 15, T. 16 N., R. 7 E., near Happy Camp, that was surveyed for a mining patent, but was not patented. Later parts of it were patented as agricultural land. They are held by Lee Effman, Mike Effman, Frank Lowrey, of Happy Camp, and various other persons. C. E. Reagan and C. T. Bonney of Happy Camp hold unpatented mining claims. They believe that there is good dredging land here, also gravel that can be worked by the hydraulic method with water from Elk Creek.

Lanky Bob and Slim Jim Claims on Whites Gulch, five miles east of Sawyers Bar, have been relocated as Gold Bug No. 1 and No. 2; and are owned by G. T. Salsbury of Etna. They are leased to Joseph T. Sugars, H. W. A. Docker, Harry Summers and Frank Cantrill of Sawyers Bar. These men, with the help of other miners in the vicinity and of the U. S. Forest Service, have just completed two miles of new road reaching the Lanky Bob mill.

The mill has been reconditioned and now consists of two 1000-lb. stamps, 5-inch by 7-inch crusher, Ellis Independence mill for regrinding, and a 4-ft. by 9-ft. plate for amalgamation below the stamps. Below this plate is a small classifier of the cone type, from which the slime goes to three home-made flotation cells containing mechanical agitators, and the sand goes to the Ellis mill, below which is an 18-inch by 2-ft. amalgamating plate. A Pelton wheel supplied with one cubic foot per second of water under a 210-ft. head drives the mill. Due to shortage of water at time of visit, an 80-hp. automobile engine had just been substituted.

Roughly 250 ft. higher than the mill is the lowest tunnel, 900 ft. long, of which 50 ft. is new work. It exposes a vein from half a foot to four feet wide, containing quartz, gouge, free gold, pyrite, and a little galena. The strike is S. 20° W. dip is 60° or 70° to the east. Walls are of an andesitic rock. According to Sugars, the quartz alone assays $40 per ton in gold. Workings on older levels above this one have caved. References to reports on this earlier work are contained in the accompanying table of mines.

Little Crumpton (Placer) is on a terrace of Klamath River in Sec. 1, T. 16 N., R. 7 E., H. M., near Happy Camp, and is owned by Clifford and Leonard Crumpton, Happy Camp. In 1934 it was leased to W. A. Minner, Bakersfield. Equipment was installed for hydraulicking with water pumped from the river to the terrace, 120 ft. above. An 8-inch centrifugal pump, rated at 2200 gallons per minute against a 340-ft. head was driven by a Liberty airplane engine by direct connection of the shafts of the two machines. The pump worked against a head of 240 ft. by gage, of which 120 ft. was lift, the balance nozzle pressure. The pump was connected to a No. 2 giant (4-inch nozzle opening) with 300 ft. of 11-inch hydraulic pipe. Gravel was sluiced through eight boxes, 12 ft. long and 20 inches wide; and material that would pass through a ½-inch screen was treated on undercurrents, 24 ft. long and 12 inches wide. About 2000 cubic yards were moved, leaving a 20 ft. bank of gravel at the back of the pit. Fuel-consumption of the engine was 10 or 11 gallons of gasoline per hour. Property was idle at time of visit in May, 1934. Equipment included a ferry across Klamath River, consisting of cable and barge.

Lowden Placer (Ariel Lowden) is in Secs. 11, 12, 14, T. 46 N., R. 12 W,. near Seiad. When this property was visited in June, 1933, some new hydraulic pipe had just been installed. According to J. H. Ladd of Seiad, who was living at the property, this supplies the mine with 1200 miner's inches of water under 100 ft. of head, from a 2½-mile ditch out of Seiad Creek. He said also that a new bedrock-ditch was to be cut immediately so that it would not be necessary to stack tailings.

T. L. Park of Seiad called at the Redding office of this division in September, 1934, and stated that he had acquired this property and that of Ladd, and had located three claims to give him a total acreage of 452½. He planned to operate under the name of *Seiad Placer Mines,*

Hydraulic pit at Lowden mine of Seiad Placer Mines, Inc.

a copartnership. References to reports on earlier operations at this mine will be found in the accompanying table of mines.

In May, 1935, E. E. Chester was finishing up a season's operations of this property, under a lease and option obtained from Seiad Placer Mines, Inc. He had three No. 3 giants, of which only two were operated at a time. The pit was on a part of the property known as the Williams channel, and exposed a bank of stratified sand and gravel 72 ft. in height. The pit is drained by a cut through the gravel and a rim of hard, dioritic bedrock, to Klamath River. The cut is 500 ft. long and has a maximum depth of 100 feet. Twelve men were at work, and kept the mine in operation for three shifts per day. At time of visit, the yardage moved during the season had not yet been measured, and the final cleanup of the sluices had not yet been made.

Lowden Placer (Jack Lowden) comprises 33 acres patented land and 40 acres unpatented mining claims in Sec. 13, T. 46 N., R. 12 W., just across Klamath River from Seiad. It has recently been purchased by S. R. Huey of Seiad and Newcastle, Pennsylvania. One giant was operated for two shifts per day, during the water-season of 1934–35, under S. K. Wood of Seiad. Equipment includes a ditch from Walker Creek to give 2000 miner's inches under a head of 200 ft. at the mine. This amount of water is available during the winter season. Wood states that a water right is owned to the first 1000 inches from Walker Creek, dating back to 1878. There are 1000 ft. of 10-inch and 15-inch pipe, a No. 2 giant, and a No. 3 giant, using nozzles with openings of 4½ inches and 6 inches respectively. Prospecting done since the water-season closed has shown that there is some very good gravel at a different location in the property from that recently mined. The good values are stated to occur in depths of gravel up to 25 feet.

Mat-a-pan Mine in Sec. 17, T. 45 N., R. 10 W., M. D. M., a half mile due west of Scott Bar, is a 20-acre claim owned by M. O. Payne of Scott Bar. In 1933, it was leased to F. E. Sette. The deposit is quite similar to that at the Quartz Hill mine at Scott Bar, which has been described in reports of this division as noted in the accompanying table of mines. It consists of a network of quartz veins in Abrams schist. Individual veins are usually not over 2 ft. wide, and for the most part only a few inches wide. The maximum width of quartz found in such a vein has been 8 ft. The white quartz carries pyrite, galena, and free gold. Ocher seams carry oxides of iron and manganese, and contain pockets that have been productive of free gold. Seven men were at work hydraulicking the decomposed surface material, also for some distance into the firmer rock below, and searching for pockets in the seams. A No. 2 giant is supplied with water through half a mile of ditch, at a head of 275 feet. In June, the water supply was such that it was necessary to let a reservoir fill for two hours to get a run of 1½ hours with the giant. Water season is eight months, lasting throughout the summer.

Minerals Recovery Corporation, Ltd., installed some machinery at the Portuguese Hydraulic Mine (see accompanying table of mines), about 1931, but has not operated there recently.

Morrison & Carlock Mine is in Sec. 13, T. 43 N., R. 10 W., four miles northwest of Greenview. Since Dec. 1, 1933, it has been operated by *Northern California Gold, Inc.*, A. C. Sloss, President, Carl Blum, Secretary, 228 Burke Building, Seattle, Washington. A reorganization under the name, *Northern California Goldfields*, is to be made. At the mine, Wm. C. Madge is in charge of the crew of 15 men.

The property is developed by a shaft, which has been driven on a fault for a distance of 225 ft. at an angle of 62° with the horizontal. One segment of the flat vein is left behind at the 50-ft. point in the shaft, then there is a 72-ft. displacement, and the shaft passes into the footwall of the second segment. At the bottom, a crosscut has been driven to this segment, which is developed to lower levels by a shaft following the dip of the vein at an average angle of 26°. This shaft is 540 ft. long on the dip of the vein, and the bottom level is called the eleventh. At time of visit in April, 1935, 50 ft. of new sinking and 300 ft. of new drifting had been done. The quartz vein is 12 inches to 14 inches in width. In stoping, the hanging-wall waste is broken for a width of about three feet to give room to work, then the quartz is broken separately. As much as possible of the waste is stored in the stopes. Country rock in the underground workings is greenstone. On the surface, an acidic dike is evident, and this is stated to have yielded considerable gold when the mine was first worked, years ago.

Ore is treated at the mine in the 10-stamp mill of 850-lb. stamps, the feed for which is broken by a 9-inch by 12-inch jaw crusher. Treatment is by amalgamation only, by which operators claim a 95% extraction. Capacity is 20 tons in 24 hours. Other equipment includes a two-cylinder air-hoist handling a one-ton skip in the main shaft; a 15-hp. electric hoist underground; 12-inch by 10-inch air compressor driven by a 60-hp. motor, and two centrifugal pumps, one driven by a 20-hp. motor, the other by a 35-hp. motor. For references, see accompanying table of mines.

Mount Vernon Mine is now assessed to Kathryn J. Pfeiffer, c/o K. K. Ash, Yreka. It comprises 70 acres of patented and 40 acres of unpatented land in Sec. 26, T. 45 N., R. 8 W. The following is quoted from State Mineralogist's Report XXVII, which also contains a sketch-map of the main level, made at that time:

"The mine is eight miles west of Yreka by a good road at an elevation of 4434 feet. It has been worked irregularly for more than 50 years. A syndicate with Kenneth K. Ash of Yreka as manager has recently done 915 ft. of new development work on the fifth or mill level, which is 1000 ft. below the outcrop on the dip of the vein. This level (Fig. 4) now consists of a 650-ft. crosscut to the vein, 900 ft. of drifting on a vein with strike S. 10° E., dip 41° E. at face, 130 ft. of drifting on a footwall branch vein, strike S. 16° W., dip the same as the other, and a 40-ft. crosscut between these. The crosscut exposes a stringer zone between the two veins. Ash estimates that a width of 8 ft. of this ran $8 per ton on the mill test. The veins are on fault fractures, and show a width of from 0 to 2 ft. of ribbon quartz, which contains a little pyrite, galena and sphalerite. The quartz is fractured and broken, indicating movement on the old plane of weakness since the veins have been formed. Ash thinks that the Mount Vernon vein, worked on the old levels, has not yet been found on this level. An old raise starts from the main drift on the fifth level and runs 550 ft. to a point above the third level, from which point there is a crosscut 140 ft. long to the Mount Vernon vein. Caving conditions make it impossible to survey these workings. From the fifth level the operators have recently sunk a 95-ft. winze, which was full of water to a point 20 ft. below the level at the time of visit. Ash states that a sample from the 30-ft. level in the winze, width 12 inches, assayed $97 per ton. According to Ash, in 1928–29, a mill run of 775 tons was made on material that was diluted 5 to 1 with country rock, which, however, may carry a little gold near the vein. Recovery was $4.65 per ton in free gold; and concentrates were saved, which are estimated at 6 tons running $97 per ton. No concentrates were obtained from the first 265 tons, which were milled before the concentrator was installed. A No. 6 Wilfley table has been added to the 5-stamp mill; and other new equipment includes a two-drill Sullivan single-stage compressor, with 40-hp. motor. On the winze are a 5-hp. geared hoist, 3-hp. motor and blower, and a self-dumping skip of 1500 lbs. capacity. The power line of the California-Oregon Power Co., running to Fort Jones, crosses the property.

South of the main workings of the Mount Vernon, a specimen was taken from an open cut on a 20 to 30-ft. dike. In the hand specimen it is a light-colored gray rock with a suggestion of a cream color, but darker. A few phenocrysts of quartz and somewhat more numerous needles of a highly altered mineral, probably hornblende, are seen in a very fine-grained ground mass. Maximum diameter of the quartz phenocrysts is 1/16 inch and maximum length of the dark needles is less than ⅛ inch. Under the microscope, large phenocrysts of feldspar are fairly abundant; but they are almost entirely altered to sericite. An occasional crystal of quartz or bleached biotite is seen. The rock is highly silicified, the groundmass consisting now of finely crystalline quartz of secondary origin and microscopic flakes of micaceous material. Much of the iron has been leached out of the rock. The needles, thought to be hornblende, are very indistinct under the microscope, because they have been entirely altered to secondary minerals. A few of the feldspar crystals show unaltered parts that can be recognized as albite. The original nature of the rock is uncertain; probably it was rhyolitic, perhaps dacitic. The type of alteration is interesting, as it indicates that mineralizing solutions have been acting on this dike. This dike is intrusive into the andesitic greenstone."

The mine has been operated continuously by Ash since the above report was written. His present crew is four men, and he mines and mills an average of four tons of ore per day. The winze has been deepened to the 130-ft. point, and two levels have been run from it. On the 77-ft. level, 200 ft. of drifting has been done, and on the 125-ft. level, 225 ft. There is also a 55-ft. crosscut into the hanging wall on the 125-level, and this entire width of country rock is stated to give assays in gold. Gross production from these workings from the winze, including stoping, is stated to have been $18,000. The ore-shoot is said to be longer as depth is attained in the winze, the length on the 125-level being 175 ft. The main level has been extended an additional 400 ft., and three stopes, 55 ft., 40 ft., and 20 ft. long, have been mined out to an average height of 50 ft. Drifting from the winze has not been extended far enough to reach these shoots. Stoping-width averages 18 inches.

To make a mill-test of the country rock, 10 tons were mined at a point where the vein had already been removed. The returns figured $7 per ton with gold at $35 per ounce, according to Ash. In addition to the minerals named in the report quoted above, some of the ore recently mined contains arsenopyrite. References to earlier reports on this mine are given in the accompanying table of mines.

Nebraska Mine comprises lots 2 and 3, NE. ¼ SE. ¼ Sec. 7, T. 46 N., R. 6 W., 122 acres of patented land. By trail, it is a mile west of the Pacific highway, on Rider Gulch, which enters the Klamath at Sawyer's Camp. A dike, probably andesitic, intrudes the graphitic schist series. It is 50 ft. or more wide. The fresh rock shows a sprinkling of minute crystals of pyrrhotite. Quartz veinlets, an inch or two wide, in the dike carry free gold. Where one of these comes in contact with the graphitic schist wall-rock, a good pocket is likely to be found. In May, 1935, two men, Harry Bieglow and Andrew Milne, were working on these veinlets of quartz. Some sampling has been done in an effort to determine if the entire dike contains enough gold for profitable mining. Apparently this sampling has not been very extensive, and is not conclusive. F. C. Dechaine holds 40 acres on the same dike, to the south of the Nebraska; and G. A. Reichman of Fort Jones holds 40 acres to the north. R. S. Stryker of Yreka holds a lease on the Nebraska.

New York Mine in Sec. 2, T. 44 N., R. 9 W., M. D. M., near Fort Jones, comprises 109 acres of patented mining claims, owned by Geo. A. Milne of Fort Jones. In 1933, G. C. Berker, Fort Jones, and C. C. Plumb, Providence, R. I., were working the mine under lease and bond.

A quartz-filled fissure vein carrying pyrite and free gold occurs in greenstone (probably Copley meta-andesite). It is developed by means of an incline-shaft, which is a crosscut, passing through the vein some distance above the 400-level. This level is 600 ft. below the surface, measured on the slope of the shaft, which has an average angle from the horizontal of 40°. The vein strikes north and south, and has a steep, variable dip to the east, averaging about 85°. On the 400-level, a drift on the vein was being driven to the north at a point 650 ft. north of the shaft. Of this distance, 400 ft. was new work. At a point 300 ft. south of the face, a fault was encountered, which displaced the north segment of the vein 40 ft. to the east. The stope that was being mined at time of visit was on this north segment, and was 114 ft. long, 20 ft. high and 3 ft. wide. Vein material was a banded, white quartz carrying free gold, pyrite and chalcopyrite. Some new work had been done to the south on the 400; but the vein was faulted and no ore-shoot was found.

On the 300-level, a drift was being driven to the north, and it was expected to cut the fault mentioned above very soon. At the south end of the 300-level, a raise was being driven for a second exit. This will connect with the 100, an adit level, which has caved, with the exception of the portion from the raise to the portal. There is also an incline from this level to the surface. It was planned to reopen the 200-level, which had caved, if the ore-shoot mentioned was found to extend above the 300-level.

Equipment included a 50-hp. electric hoist; Sullivan angle compound compressor, 12 inches by 7½ inches by 8 inches, driven by 75-hp. motor; 6-inch Byron Jackson turbine pump; No. 5 Cameron sinking pump; Sullivan drill-sharpener with oil-burning furnace for heating drills. Ore was hauled to the Golden Eagle mill, about a mile distant, with a 2½-ton Ford truck. The mill contains 16 stamps of a capacity of 35 tons per day, plate-amalgamation, and a Wilfley table. Crushing was to 35-mesh. At time of visit, 22 men were employed. For references to earlier reports on New York mine and Golden Eagle mill, see accompanying table of mines.

Nigger Boy Mine comprises 240 acres of patented land, former railroad land, in Sec. 1, T. 46 N., R. 7 W. It is owned by O. K. Wilson, San Diego, and is optioned by Lee DuBois, Yreka. From the Klamath River highway, a good road has recently been built up Ash Creek to this mine and beyond, by the Civilian Conservation Corps.

Old adit levels of a total length of 3700 ft. expose a flat vein at elevations varying from 135 ft. above Ash Creek to 650 ft. higher. The country rock is a metamorphosed sedimentary series now appearing mostly as a graphitic schist. The average dip of the veins is under 10°, and widths average about 2½ ft. with a 5-ft. maximum. The main vein is of white ribbon-quartz carrying free gold and auriferous pyrite. Some of the vein filling is white calcite, and this often carries free gold. Apparently the occurrence is somewhat similar to that at the Hazel or Jillson mine, a little over a mile to the northeast. At the Nigger Boy, roughly perhaps 50,000 or 75,000 tons of ore have been mined and milled, said to run about $30 per ton. This work was done in the years from 1891 to 1909, and from 1911 to 1915. Equipment, including a 6-stamp mill of 1000-lb. stamps, two compressors, track, cars, etc., was removed in 1921.

DuBois has been prospecting surface material with cuts and short adits. He finds that some of the country rock containing quartz stringers assays from $3 to $5 per ton. References to reports on earlier work at this mine are contained in the accompanying table of mines.

Norcal Mining Company is to be incorporated by E. G. Sundfelt, R. S. Fox and E. G. Lindstrom of Sawyers Bar, California, and Seattle, Washington. They are opening up a group of old mines in T. 39 N., R. 11 W., near Rollin, six miles south of Sawyers Bar. Included in the group are the Ida May, Klamath, Union-Central, Spears, and Wescoatt or Boss, a total of 18 or 20 claims and two or three mill sites. An effort is being made to block out sufficient tonnage in these mines to justify the construction of a central milling plant.

At the Central, a portable compressor has been installed, and 170 ft. of new workings have been driven; also 1100 ft. of old caved tunnels have been cleaned out. At the Ida May, 600 ft. of old tunnels have been cleaned out, and 100 ft. of new workings driven. At the Klamath, 600 ft. of old workings have been reopened. This work was being continued with a crew varying from 15 to 30 men at the time of visit in June, 1935.

A two-stamp mill has been put in working order for use as a pilot mill. It contains an ore-bin, crusher, and a Denver Sub-A-unit flotation cell. At time of visit, the fifth mill-test of 60 tons was being run. For power, a new hydro-electric generator developing 60 hp. to 70 hp. is in use.

Ore is found at the Union, Central and Klamath in a flat vein of banded quartz with widths of 6 ft. to 8 ft., dipping on the average 18° easterly. Two other veins of the same character and nearly parallel to the one mentioned are found. In places they are said to unite to form a large orebody. At the Ida May, farther down the mountain, a vein similar in character, but with the quartz crushed to small fragments, and with a large amount of gouge, is found. It is thought to be a faulted segment of one of the veins above. Country rock is graphitic schist; and a light-colored dike, possibly andesitic, is associated with the vein. This is locally called 'cab'. References to reports on former operations at these mines are contained in the accompanying table of mines.

Norma Mine is in Sec. 28, 29, T. 46 N., R. 8 W. A total of 240 acres of unpatented mining claims is held by Geo. J. Cottle of Walker post office, as follows: Norma mine, 3 claims; Royal, 4 claims; Edna, 3 claims; San Jose, 2 claims. He has been prospecting small quartz veins, a few inches wide, in diorite. Some of the work exposes small veins, a few inches wide, of heavy iron sulphide, probably carrying some copper, in the diorite. Several hundred feet of open cuts and adits have been driven.

O'Connell Gold Mines Incorporated is reported to have done some development work at the Homestake and Hogan mines, 12 miles northeast of Sawyers Bar; but these mines were idle when field work was done in that district, and they were not visited. References to earlier reports on them are given in the accompanying table of mines.

Oro Grande Mining Co. The following is quoted from State Mineralogist's Report XXVII, chapter for January, 1931:

"The McKeen mine of this company is in Sec. 36, T. 40 N., R. 9 W., near Callahan or 31 miles from the railroad at Gazelle over a good road. Hugh McKinnie, Callahan, California, is president and R. D. McKinnie is secretary. The property comprises 480 acres of patented land at an elevation of 4000 ft. Plenty of water from Boulder Creek, and good pine, fir and cedar timber are available on the property. It was discovered in 1881; and was last worked in 1929. Four veins are known; but only one has been developed. This is a fissure vein of white quartz with pyrite. The pyrite forms about 10% of the vein filling, and the gold is in pyrite. The vein, of an average width of 2 ft. and a maximum of 5 ft., strikes S. 35° W., and dips 85° to the southwest. On the lowest or No. 1 tunnel level, 2825 ft. of drifting have been done; and on the No. 2 level, 220 ft. above, 2300 ft. of drifting. A raise connects the two levels at a point 1500 ft. from the portal of No. 1 tunnel. Levels above No. 2 are stoped out and caved. Between No. 1 and 2 levels, owners claim ore reserves of 30,000 tons of ore of a value of $8 per ton. They state that three pay shoots have a length of 250 ft. each, and that the No. 1 tunnel has been in a shoot for the last 75 ft. The vein has a gouge parting on both walls and breaks clean of country rock without difficulty in the shrinkage stopes.

Equipment includes a compressor, capacity 500 cu. ft. of free air per minute, machine drills for both stoping and drifting, cars and track. In the mill are a Blake crusher, Hardinge ball mill, Dorr classifier, two 10-ft. amalgamating plates, K & K flotation machine. The capacity is 35 tons through 60 mesh in 24 hours, with an extraction of 90%. A laboratory test with cyanide is stated by Hugh McKinnie to have an extraction of 94 to 96%. He says that a production of $250,000 is recorded and that probably an equal amount additional has been produced but not recorded. The only recent milling was a test run of 10 days in 1929. Water power from Boulder Creek is used to drive the machinery. A mile of ditch and 1000 feet of 14-inch pipe supply 500 miner's inches with a 260-ft. head.

Numerous cuts and small tunnels were observed on quartz veins on the slope of the hill to the northwest of the mine. As one climbs this hill the float changes from granodiorite, which forms the walls of the McKeen vein, to serpentine, showing that the deposit is very close to the contact. Pieces of float of white, aplitic-appearing material, with only scattered crystals of biotite, are common in places on the slope also."

References to earlier reports on this mine are given in the accompanying table of mines.

Paradise Group comprises 460 acres in Sec. 18, T. 15 N., R. 7 E., H. M., 10 miles southwest of Happy Camp, held by the estate of Reeves Davis, W. F. Davis, executor, 427 J Street, Sacramento. Three of the claims are patented under M. S. 5155, named San Francisco, Hunter's Paradise and Jolly Joker. Some gossan is found here; also larger bodies of schist mineralized with quartz seams and sulphides of iron and copper. Practically all of the surface material is heavily stained by oxides of iron derived from these sulphides. According to W. F. Davis, the deposit was to be extensively drilled and sampled, in 1934, with the idea that it may contain enough gold to pay by mining on a large scale with power-shovels.

Quartz Hill Mine. The following is quoted from State Mineralogist's Report XXVII, chapter for January, 1931, which contains also photographs of the mine and a sketch-map of the deposit:

"It is a property of about 75 acres of patented ground in Sec. 16, T. 45 N., R. 10 W., at the junction of Mill Creek and Scott River, near the town of Scott Bar. By road it is about 45 miles from either Yreka, the county seat of Siskiyou county, or Hornbrook on the main line of the Southern Pacific railroad from San Francisco to Portland. Thirty miles of this road are graded highway following the Klamath River; and the balance is fairly good. Patents are covered by Mineral Surveys, Lot 37, Marfield & Co. Placer Mine, and California Placer Mine, M. S. 3493. The owner is Harry G. Noonan, 2801 Jackson St., San Francisco. In 1860, or earlier, the mine was started as a hydraulic mine working on material eroded from the large system of quartz veins now exposed. A small mill of 10 stamps of 750 lbs. each was used in some of this early work, with amalgamation for the recovery of the gold. Present operators have worked the deposit for 25 years. Three or four men blast the hard quartz and schist and then hydraulic this broken material. Giants, discharging a large volume of water at a high pressure, wash the broken rock through a regular sluice with riffles in the bottom to catch the gold. The face of the pit is examined after each blast; and if any high grade is found it is sorted out and crushed in a small ball mill, run by a 5-hp. air motor, and equipped with amalgamating plates.

Water is taken from Scott River by means of a ditch, seven miles long, with a capacity of 1500 miner's inches, and a head at the mine of 225 feet. There is no interruption in the supply of water in the dry season. A giant with a 5-inch nozzle is used in the pit, one with 4½-inch nozzle at the head of the sluice, and one with 4¾-inch nozzle to wash the tailings down the river. Present operations are roughly 25 or 30 ft. higher in elevation than the river. For drilling with the jackhammers, an 8-in. by 9-in. Ingersoll compressor driven by water power furnishes the air."

During 1935, Geo. A. Milne of Fort Jones, and George Noonan of Scott Bar have been leasing the property and working it by the same hydraulic method described. References to reports on earlier operations are contained in the accompanying table of gold mines.

Queen of Sheba Mine includes the old Eton mine and enough other patented land to make up a total of about 250 acres, in Sec. 10, T. 40 N., R. 9 W., M. D. M., between Etna and Callahan. It is being developed extensively by Goldberg & Sullivan, c/o Abe Goldberg, Etna. The method is the same as that used at the King Solomon. At time of visit, late in 1934, half a mile of trenches, 12 ft. wide, and as much as 20 ft. deep, had been excavated with a large diesel Caterpillar tractor with blade in front (bulldozer). These trenches were scattered over an area of 50 acres, and were designed to prospect a metamorphic zone, 600 ft. wide, in schistose sediments between serpentine and granodiorite. The contact-metamorphic zone contains garnets, sulphides of iron, and free gold. From the trenches, 1500 samples had been run through a new assay office at the mine. Each sample weighed 80 lbs., and was crushed and quartered to yield 4 lbs. for a test in an amalgamating

Queen of Sheba. Pilot mill near Etna.

muller. Tailing from each amalgamation test was assayed by the regular fire-assay. The oxidized zone is only a few feet deep here; and trenches deeper than that are excavated in fairly hard rock requiring blasting.

A new test mill of a capacity of 30 tons in 24 hours of the soft, surface ore had just been built. It included a 125-ton coarse-ore bin, 7-inch by 10-inch crusher, 100-ton fine-ore bin, belt conveyor to double-deck vibrating screen, 3-mesh and 30-mesh. Undersize was treated in Gibson impact amalgamator. Oversize was ground in a 4-ft. by 4-ft. Union Iron Works ball-mill with open discharge, and bucket-elevator to deliver product to vibrating screen already mentioned, with which it was in closed circuit. Two Ford V-8 truck engines furnished the power. Other equipment included a Thew gasoline shovel with a ¾-yard dipper, a road plow, and a dump-truck. New flume of a length of 2700 feet had been built in the ditch-line from Sugar Creek to furnish 30 miner's inches of water to the mill.

Reeve Ranch Mine (placer) in Sec. 2, 11, T. 16 N., R. 7 E., H. M., two miles east of Happy Camp, is assessed to Miss M. A. Reeve, 729 Oak Street, San Francisco. It contains 211 acres, patented, and 70 acres of river channel. James T. Logan of Happy Camp is leasing the property, and is working it by the hydraulic method. The deposit consists of a 20-ft. depth of gravel on a low terrace of Klamath River, just above the present stream. Logan uses a hydraulic elevator to stack the tailing. Water supply comes from Indian Creek through 2¼ miles of ditch carrying 30 cu. ft. per second, and giving a head of 80 ft. at the mine. Pipe installation consists of 2500 ft. of 20-inch and 30-inch pipe to supply the hydraulic elevator (Joshua Hendy) and one No. 2 giant with 3-inch nozzle opening; also 2000 ft. of 11-inch and 13-inch pipe to supply two No. 2 giants with 3-inch nozzles. The hydraulic elevator raises gravel 14 ft. vertically to a sluice 42 inches wide by 24 inches deep by 108 ft. long with a grade of 3¼ inches to 12 feet. It has a 10-inch throat, and will lift a rock of that size. Water is ponded behind a timber dam, and part of it is returned to the elevator. This saves about three cubic feet of water per second, and keeps much of the silt out of the river. The sluice is rolled ahead on rollers of 6-inch pipe. In the spring of 1934, Logan was mining a 16-ft. depth of gravel, and stated that he was handling 1000 cu. yd. per 24 hours of gravel running 30¢ to 50¢ per cubic yard in gold (gold at $20) with a crew of seven to eight men. References to reports on earlier mining here are given in the accompanying table of mines.

Reno Group of two unpatented claims in Sec. 14, T. 44 N., R. 9 W., M. D. M., near Fort Jones, was being prospected in the fall of 1933 by the owner, N. J. Rowley of Fort Jones. A mile of new road had been built to the property. An andesitic dike intrudes the greenstone here. In the dike are stringers and bunches of quartz, a fraction of an inch wide up to an inch or two wide, also considerable pyrite in the fresher rock. The pyrite is oxidized in the surface material, coloring it

Reeve Ranch mine near Happy Camp. Hydraulic elevator.

brown. Both oxidized and unoxidized dike rock are stated to give pannings in free gold. Development work consisted of open cuts and a 10-ft. shaft. Four men were at work.

On patented ground in the adjoining SE. ¼ Sec. 11, a 50-ft. crosscut adit runs N. 70° E. to strike a fracture zone dipping 45° to the east. It contains a mixture of gouge and quartz for a width of 3 ft. to 4 ft.

Short drifts, a raise and a winze were run on this years ago; and a small production of gold is said to have resulted.

Richardson Placer (DelNorte Mining Co.) is an old hydraulic mine in Sec. 1, T. 16 N., R. 7 E., near Happy Camp, on the opposite side of Klamath River. It is controlled by the Richardson Estate, W. W. Williams, Administrator, 486 California St., San Francisco, California. Of the 450 acres of patented ground, 50 acres were worked off years ago in a strip averaging 600 ft. wide. The banks left at the faces of these old hydraulic pits are 30 feet high, of which about a third appears to be overburden. Tunnels have been driven 200 and 300 ft. into these banks, and are entirely in gravel, which probably extends considerably farther. Local miners state that this gravel is of a commercial grade at the present price of gold ($35). To determine the grade accurately will require an extensive sampling campaign. A good water-right is said to be available at Elk Creek.

Riverside Placer is a claim of 20 acres in Sec. 32, T. 17 N., R. 8 E., H. M., held by E. D. Friend of Happy Camp. Two men were drifting and ground-sluicing in 1934. It has not been visited.

Roff Mine (Cory or Bridger mine) comprises one patented and three unpatented claims in Sec. 34, T. 41 N., R. 9 W., M. D. M., held by V. W. (Dan) Roff of Etna. An 18-inch vein of white quartz stained with oxide of iron, and carrying free gold is found in a metamorphosed sedimentary series of slates and cherts. Workings examined by the writer include an 18-ft. shaft and a 150-ft. drift adit at an elevation 100 ft. lower. In the shaft a 2-ft. width of quartz was exposed, which Roff stated would mill $9 (gold at $20) in free gold per ton. From the tunnel, the vein was stoped 40 years ago, when the property was known as the Cory mine. A cave in the face was being cleaned out; and Roff expected to drive the tunnel ahead for 15 or 20 feet to get under the ore found in the shaft. Roff describes other workings, all thought to be on the same vein as follows: At an elevation 40 ft. below the 150-ft. tunnel, the vein is exposed in a 100-ft. crosscut tunnel with 6 ft. of drifting, showing a width of 2 ft. At an elevation 120 ft. still lower is a 320-ft. tunnel that is thought to parallel the vein; and crosscuts might pick it up. On the next claim to the south is a 120-ft. tunnel, of which 100 ft. is drifting with the vein stoped out to the surface, 60 ft. above. On the second claim to the south is a 520-ft. crosscut adit that did not cut the vein on account of faulting. There is another adit at the south end of the group; but the portal has caved. The vein strikes northerly and southerly; and the dip is nearly vertical.

Ore from the shaft was being hauled to a mill about half a mile below, with a capacity of 8 tons per 24 hours. It contained a 7-inch by 9-inch crusher, Ellis mill, and plates for amalgamation, 4 ft. by 10 ft. Power was furnished by a Chevrolet automobile engine. There was also a 36-inch by 40-inch ball-mill, which had not been installed.

In Sec. 27 to the north, Roff holds a claim on a 50-ft. dike, probably andesitic, which he says averages $1.50 (gold at $20) per ton in gold. In a 50-ft. adit giving a depth of 30 ft., he says the average is 80¢, and for the last 10 ft. is $1. The dike can be traced for 10 miles, passing through the Eton claims, according to Roff.

Salmon River Mines Co., Inc., E. C. Latchem, President, V. W. Peterson, Secretary, Callahan, California, is developing the Foster mine in Sec. 12 (?), T. 39 N., R. 10 W. It is reached by six miles of mountain trail from the end of the road up South Fork of Scott River near Callahan. Elevation is 6600 ft. A small mill was operated on the property years ago. New work has been done in a 950-ft. adit to pick up ore on the dip as shown by a survey of old workings. The adit is low enough to give 200 ft. of new backs on the vein. At time of visit, the face exposed a 5-ft. quartz stringer zone with vertical dip. According to E. C. Wood, who was in charge, a sample taken across the face assayed $12 per ton in gold. The footwall formation is a graphitic schist similar to that at Sawyers Bar; the hanging wall probably schist of a different kind. Three men have been at work. Equipment includes a compressor driven by a gasoline engine.

Schroeder Mine. The following is quoted from State Mineralogist's Report XXVII, chapter for January, 1931:

"This mine in Sec. 17, T. 45 N., R. 8 W., is assessed to M. C. Beem and others of Fort Jones; but it has recently been reported sold to the Fidelity Metals Corporation. V. O. Bartoo is in charge, with I. B. Scrimger as mining engineer. The mine address is Fort Jones. The property is seven miles in an air line west of Yreka, and is reached over the old road to Fort Jones, which crosses a summit near the Mt. Vernon mine at an elevation of more than 5000 feet. Turning from this road at Deadwood Creek, the road to the mine starts climbing the mountain, and is very steep and winding for 3½ miles. In 1895, at least five veins, or possibly faulted segments of veins, had been exposed on this property; and thousands of feet of development work had been done from adits at several different elevations on the steep side of the mountain. Ore was being treated in a ten-stamp mill operated by steam power at the rate of 12 to 13 tons per 12-hour shift. In 1927, the *Sterling Gold Mining Co.* did some work here; but very little ore was produced.

At the time of visit, an old tunnel on a vein called the Snowflake had been cleaned out and retimbered, so that the vein was exposed for a length of 100 feet. A width of 7 to 8 feet of shattered quartz heavily stained with iron oxides, has a strike of N. 80° E., and dip of 60° to 75° to the south. The operators stated that this gives an average assay of $8.80 per ton in free gold. The mill tunnel, called the 1600, is at an elevation 350 ft. below, measured vertically. The name '1600' comes from early-day measurements taken on the surface of the ground from some higher adit. When visited, this main level was 2000 ft. long, and was being driven ahead on a course of S. 45° W. to cut the Snowflake vein. The tunnel had just been passed through a quartz stringer zone, which carried some gold; and work was to be done on this later. A compressor driven by power from the lines of the California-Oregon Power Co. was being used on this work. The 10-stamp mill was not operating.

Country rock is largely greenstone which has been highly altered. Some fine-grained granitic-appearing rock with crystals only about a millimeter in diameter was observed near the office, and some of possibly the same much silicified near the portal of the main tunnel. A dioritic intrusion was observed at a point higher on the mountain. The geology is somewhat complicated here; and apparently a complete geological examination would be of great benefit. This will require first the preparation of a good map on a large scale, based on a transit survey, of the surface and such workings as are open. With surface and underground exposures of veins, faults and formations carefully plotted and projected on such a map, a much better understanding of the vein system would result."

The mine is now owned by *Fidelity Metals Corporation*, Frank E. Jones, President, 1403 Santa Fe Ave., Los Angeles. This company did about 1000 ft. of new work on the lowest (1600) level, mostly in the fine-grained greenstone. Apparently this work is in the footwall of the main vein system; but the presence of faulting, and the distance of this level below the old productive levels (300 ft. measured vertically) make the exact position of this main vein system uncertain.

In the report quoted above, a quartz stringer zone is mentioned. This was later called the East & West vein, and roughly 200 ft. of drifting was done on it. A stope 60 ft. long, 60 ft. high and 5 ft. wide was made also. Some very good ore was found, but it was not as wide as the stope, and was diluted considerably by wall-rock in mining.

Although the mine was idle when visited, C. H. Van Nest, Box 275, Yreka, who was in charge, stated that the mill may be started to treat ore from the new discovery at the Katie May mine, 6½ miles distant. They are both near a good road running along the top of the ridge, 5000 to 6000 ft. in elevation, which continues on fairly easy grades to Yreka. Treatment in the mill is amalgamation on plates only. References to earlier reports on this mine are contained in the accompanying table of mines.

Sheba Mine is near the Buzzard Hill mine on a deposit of the same kind. It has been described in earlier reports under the name, *Outlook*. See accompanying table of mines. Present owners are Peter Grant and Alfred Effman of Happy Camp. Annual labor to hold the claims is all that has been done recently.

Short Bend Placer of 20 acres in Sec. 18, T. 16 N., R. 8 E., H. M., is held by Forest Moore and C. G. Lewis of Happy Camp. The adjoining River Bend claim of 20 acres is held by Moore only. They cover a low bar of Klamath River, about 2500 ft. long and 150 ft. wide, also river-bed of a somewhat larger area. In the Spring of 1934, Moore and Lewis were installing a scraper of ¼-cu. yd. capacity to be pulled by a winch on a Fordson tractor. They planned to dump the scraper into a sluice with screen of ½-inch mesh in the bottom, and to treat the undersize on tables similar to dredge-tables; also to screen out sizes below 14-mesh for further recovery of gold and platinum by methods under development. Moore and Lewis stated that 100 cu. yds. of gravel shoveled into sluices during the previous summer yielded a total of $94 (gold at $20), and that the recovery was not complete. Six prospect shafts 3 ft. to 4½ ft. deep were said to have given returns of $1.50 per cubic yard (gold at $35). Water for washing is pumped from the river.

Siskiyou County Mine comprises five claims, the old Bumblebee group in Sec. 31, T. 18 N., R. 7 E., H. M., just below the Classic Hill mine. The five claims are now held by Siskiyou County (address: Board of Supervisors, Yreka). They have been leased to individuals in the past, apparently for the purpose of having annual work performed. During the past few years, while annual labor has been excused by acts of Congress, they have not been leased. John Attebury leased the property in 1929, 1930 and 1931, working one season and two or three months of another season, and producing roughly $1800. A hydraulic pit on a terrace of Indian Creek is worked with a No. 2 giant supplied with water from Mill Creek under a 200-ft. head. The ditch is a mile and a half long and carries 1500 miner's inches. Attebury stated that he was not justified in properly equipping the property because it is leased for only one year at a time, and that amount of time is needed to recondition equipment. In his work only about half of the available head was used.

Siskiyou Metals Company did some work on Crapo Creek near Forks of Salmon in 1929 and 1930. At times of several visits to the district in later years, the property was idle.

Siskiyou Mines and Development Company issued a prospectus on a mine near Callahan, from Everett and Bremerton, Washington. Mark

F. Mendenhall is president and General Manager, and Wm. P. Alvik, 1107 Hewitt Ave., Everett, is Secretary-Treasurer. Apparently the mine is the *Salsbury* at the head of Mill Creek, south of Callahan, although this name does not appear in the prospectus. The mine has not been visited.

Soda Mint Mine comprises 40 acres in Sec. 29, T. 39 N., R. 3 W., leased from *Hewitt Lumber Co.*, by J. J. Murphy of Weed. From Castle Crags there is a dirt road up Soda Creek, then a rough road up Little Soda Creek, from the end of which the property is a mile up the mountain by trail. An effort has been made, by the driving of 1100 ft. of tunnels, to develop a body of low-grade gold ore large enough to mine with power shovels. Murphy states that he has obtained assays as high as $5.50 per ton with gold figured at $20 per ounce. The deposit is said to be 200 ft. wide. It was apparently originally a fine-grained igneous rock very thoroughly impregnated with pyrite in abundant, minute crystals. There are also tiny seams and vugs filled with quartz. The mass has been so thoroughly weathered and disintegrated on the surface that it now appears as a fine-grained sand of yellowish and reddish-brown color.

Sunnyside Placer in Sec. 1, T. 16 N., R. 7 E., a few miles northeast of Happy Camp, is owned by H. J. Barton of Yreka. For hydraulicking, water supply is taken from Cade Creek through two miles of ditch, giving a head of 150 ft. at the mine. There is one No. 2 giant using a nozzle with a 3-inch opening. A depth of 12 ft. of gravel is found on a high terrace, and considerable faulting is evident in the bedrock, making it irregular to mine. A little mining is done each winter by lessees.

Thompson Creek Mine in Sec. 15, 16, T. 18 N., R. 7 E., H. M., of seven unpatented claims, is held by S. K. Wood of Seiad and others. It is nine miles by trail, up Thompson Creek, from the Klamath River. According to Wood, a 2-ft. quartz vein carrying gold has been traced on the surface for a length of 7000 ft. Strike is east and west, and dip is vertical. It is developed by two adits, one a drift 270 ft. in length, the other comprising 100 ft. of crosscut and 140 ft. of drift. The second is 140 ft. lower in elevation than the first. Eight or 10 tons of ore said to assay $70 per ton are on the dump.

Turk Mine on the edge of Quartz Valley, about 30 miles by road southwest of Yreka, comprises 160 acres in Sec. 7, T. 43 N., R. 9 W., and Sec. 12, T. 43 N., R. 10 W. It is being operated by Banks & Maginnis, Inc., with J. P. Maginnis of Yreka in charge. An upper drift-level, giving backs of 90 ft., exposes the vein for a length of 400 ft., with a strike of N. 25° W., and dip 62° NE. Operators state that this work has exposed two shoots of mill-ore, one 35 ft. long, the other 60 ft. long, with average widths of 15 ft. to 16 ft. A 300-ft. crosscut has been driven toward another known vein, the Turk vein worked years ago, but this vein has not yet been reached. At an elevation 155 ft. lower, a second level is being driven. At time of visit, in April, 1935, it had been on the vein for 450 ft. and in an ore-shoot for 30 ft. Total length was 600 ft. The face showed a width of 8 ft. of quartz, gouge, and silicified greenstone. This level is 280 ft. higher in elevation than the mill, with which it is connected by a 700-ft., three-rail gravity tram.

A 65-ton bin is provided at the top of the tram, and primary crushing is done here by a No. 2 Gates gyratory crusher driven by a 25-hp. electric motor. Crushed ore is delivered to a 100-ton bin at the mill.

The mill contains 10 stamps of 1050 lb. each, plates for amalgamation, and a Wilfley concentrating table. Capacity is 34 tons in 24 hours. Other equipment includes a 310-cu. ft. air-cooled compressor driven by a 50-hp. motor, and a small blower for ventilation. Electric power is

Turk Mine on the edge of Quartz Valley. Mill and gravity tram.

used throughout, a new line a mile and a quarter long having been built from the Morrison & Carlock mine.

The average crew comprises 15 men, of which 10 are employed in the mine. Allen Maginnis is mine foreman, and Wm. Maginnis is mill foreman and assayer. A third vein, called the Keenan, is known on the property, and a shaft has been started on it.

Underland Mine (Julia) of six claims in Sec. 28, T. 39 N., R. 3 W., is owned by J. J. Murphy of Weed, J. A. Barnett, and others. From Castle Crags, a dirt road goes up Soda Creek, then a rough road up Little Soda Creek to the mill, for a total distance of 6½ miles. The mine is a few hundred feet above on the mountain, and is reached by trail. With the mill, it is connected by an iron slide, through which the ore runs by gravity. The ore consists largely of gouge found in a seam from a foot and a half to four feet wide in slate. The bedding of the slate is practically flat, while the gouge has a dip of 30° to 40°. Several hundred feet of adits have been run at different levels, but these are now partly caved. In one of them a 3-ft. quartz vein carrying $7 per ton in gold (gold at $20) is stated to have been found. The property is now leased to J. A. Barnett and John Brewer, who are planning on running a new adit-level.

Equipment includes a two-stamp mill of 1000-lb. stamps, plates for amalgamation and a Wilfley concentrating table. Water power is used to drive this machinery, but is available for only a part of the year. Murphy states that table concentrates assay $90 per ton in gold (gold at $20 per ounce).

Victory Gold Mines Company is in Sec. 16, 17, 20, T. 40 N., R. 10 W., M. D. M., on the South Fork of Russian Creek, three miles by road from Finley Camp. John Nefroney of Etna acquired title to this property, a year or two ago; but another change of hands has recently been reported. At time of visit in August, 1934, Nefroney was personally operating a new mill of a rated capacity of 24 tons in 24 hours. Ore was ground in a rod-mill and treated by amalgamation only; although the mill contains a concentrating table of the Wilfley type. Power was furnished by a 72-kw. hydro-electric generator installed in 1932. Shortage of water for this limited operations to only a part of the day. Nefroney stated that he did not have time to show the mine. C. A. Logan described geology and development work in State Mineralogist's Report XXI, p. 459; also the Advance on p. 430. Probably no large amount of development work has been done since that time. Ore for the run of the mill mentioned above was being delivered on mule-back.

White Bear Mine was being operated in 1934 by *Mayland Mining Company* (not incorporated.) A. H. Mayland, Calgary, Alberta, Canada, and S. W. Brethorst, Hoge Building, Seattle, Washington, are officials. Arthur J. Theis, Sawyers Bar, is manager. The mine is in T. 39 N., R. 12 W., M. D. M., two miles by road south of Black Bear mine; and 12 unpatented claims are held.

A quartz fissure vein carrying free gold and auriferous sulphides occurs here in graphitic schist. A decomposed dike (aplite?) of a width of a few feet accompanies the vein which consists of bands of white quartz alternating with bands of black schist. The vein averages 4 ft. in width with a maximum of 9 ft., strikes east and west, and has a vertical dip. It is developed by three adit-levels, of which the two upper ones in the oxidized zone have been worked out. The No. 3 level consists of a 300-ft. crosscut to the vein and 700 ft. of drifting.

At time of visit, in the Fall of 1934, the mill had just been remodeled, and contained a primary crusher, 9 inches by 12 inches, secondary crusher, 8 inches by 8 inches, Straub Challenge feeder, Straub rib-cone ball mill, 4 ft. by 3 ft., Dorr-type Simplex classifier, 27 inches by 16 ft., launder with riffles, two plates for amalgamation, 4 ft. by 6 ft., two Kraut flotation cells, two American Cyanamid flotation cells, wooden settling tank for concentrates, and drying pan. Power for the mill was furnished by a 50-hp. semi-diesel engine. There were also a 360-cubic foot Chicago Pneumatic air compressor driven by a semi-diesel engine, and machine drills.

Buildings included a small machine shop, cook house, mess hall and store room in one building, assay office with power crusher and grinder, office and storeroom combined, 8-room staff house, 3 dwellings, 4 bunk houses of one room each, and a barn. Fifteen men were employed.

White Dog Mine is one mile south of the Hansen, and is held by H. W. A. Docker of Sawyers Bar.

Wingate Hill Placer in Sec. 6, T 15 N., R. 7 E., H. M., eight miles southwest of Happy Camp, is held by O. Y. Anderson and Van Hoy. According to C. E. Reagan of Happy Camp, 10,000 cubic yards of gravel moved from this deposit by hydraulic mining, during the season ending in the Spring of 1934, yielded $2800. Some of the last gravel moved is stated to have averaged $1 per cubic yard. The property has not been visited. References to reports on earlier operations here are contained in the accompanying table of mines.

Yellow Rose Mine. The following is quoted from State Mineralogist's Report XXVII, chapter for January, 1931:

"This mine in Sec. 20, T. 37 N., R. 9 W., is the property of *McCormick-Saeltzer Co.,* D. V. Saeltzer and R. A. Saeltzer of Redding, Clifford Saeltzer and Dorothea Sutton of San Francisco, and J. C. Boddicker of Carrville. It comprises the patented Yellow Rose of Texas, U. S. Mineral Survey, 3780, 14.91 acres; Yellow Rose mill site, 5 acres, and the Red Rose and Friendship claims. One can now ride to within four miles of this group over the very rough road built by the McNamara Mining Co. to the Nash mine. The last four miles are over a mountain trail. The Yellow Rose and the Dorleska are on the same system of dikes and veins, located on the divide between Salmon River and Coffee Creek, at an elevation of 6500 to 7000 feet. Main workings of the Yellow Rose are on the south side of this divide and those of the Dorleska on the north side. In 1912, MacDonald described this deposit as a serpentine mass cut by large and small lamprophyre (camptonite and kersantite) dikes. A 50-ft dike, strike N. 16° E., dip 75° W., cuts the serpentine near the schist contact, and parallel to it; and a fault zone is also parallel. MacDonald says, 'Near the Yellow Rose mine the east side of this dike contains hornblende prisms and is the "Crow's foot porphyry" of the prospector. The west side shows biotite and some quartz. * * * The ore deposits of these mines are similar. The fault zone is mineralized along the footwall of the composite dike, which is also locally mineralized where cut by side fissures. Rich shoots of clayey material which dip 45° N., characterize some of the fissure intersections.'

Harry M. Thompson of Redding, who holds an option on the Yellow Rose, has two men at work on the property. This work is being done with the idea that the 'clayey' ore mentioned by MacDonald is the result of fault drag from quartz veins in the dikes. It is thought that former operators followed the line of least resistance with development work, and kept in the soft material, leaving the dikes, which are silicified and extremely hard, as an unbroken wall on one side of the working. New development work will break into these hard dikes in an effort to find the ends of quartz veins that have been sheared off by the faulting. Thompson states that since the visit of the writer to the mine, these ideas have been proved correct; and ore has been found in one of the dikes.

At the time of visit, a tunnel about 200 ft. higher in elevation than the mill level was being opened. At the 600-ft. point in this tunnel there is a good exposure of a dike, strike due N., dip 81° W., in contact with the serpentine. The tunnel has apparently followed this contact for some distance; but it is so tightly timbered that this can not be well observed. Twenty-five feet to the east a second dike is seen in contact with the serpentine; and the gouge on this contact is said to pan gold.

Preparations were being made to follow this contact with a drift toward the Dorleska. All of the workings of the latter are closed by caving. An old 3½-ft. Huntington mill, which was driven by steam, stands on the Yellow Rose property.

Bibl: State Mineralogist's Report XXII, p. 26."

TABLE OF GOLD MINES AND PROSPECTS, SISKIYOU COUNTY

With References to State Mineralogist's Reports

Ranges east (E) refer to Humboldt Meridian.　　Ranges west (W) refer to Mount Diablo Meridian.

Name of mine	Sec.	Twp.	Range	Remarks on location or owner	Elevation, feet	Area, acres	References
Abe Lincoln Placer				S. side Klamath opposite Portuguese	1,400		XXI, p. 462
Abel W. Cook, Placer.				5 mi. above mouth Horse Creek.	1,950		XXI, p. 462
Advance	16	40 N.	10 W.	John Nefroney, Edna. E. J. Miley and Edna McEwen.	5,000	80	XXI, p. 430; XIV, p. 825; see also Victory Gold Mines
Allgood & Castell Placer	12	11 N.	7 E.?	3 mi. SE. Somes Bar		73½	XIV, p. 860
Amalgamated				See Gold and Nickel		40	
Amazon				2 mi. S. Henley	2,325		XIII, p. 386
Amlack				See Humpback			
Ames				14 mi. S. Happy Camp	1,200	20	XXI, p. 431
Anderson Placer				See Joubert			
Ascondry Drift Placer	10	46 N.	9 W.	Lawrence Litchen	2,000	100	XIV, p. 861
Bailey	35	46 N.	9 W.		5,000		VIII, p. 624; XIV, p. 826; XXI, p. 431
Baker	15	44 N.	9 W.	J. C. King, Callahan	5,000		XIV, p. 826; XXI, p. 431
Baldwin Group	18	40 N.	7 W.	M. H. Balfrey, Gazelle.	3,300	100	XXI, p. 431
Balfrey	28	41 N.	7 W.	See Ida May & California Con.	4,500	800	See text this report.
Ball				Cecilville District.			
Ball & Goforth				6 mi. N. Callahan; R. F. Ballingal, Callahan	4,000		XXI, p. 431
Ballingal Placer				See Turk			XX, p. 180; XXI, p. 464
Banks & Maginnis, Inc.				In Quartz Valley near Evans Gulch.			XII, p. 277; XIII, p. 387
Banner				Oak Bar District	2,000		XIV, p. 844; XXI, p. 464
Bark House Creek Placer	17	40 N.	7 W.	Simon Barandun, Gazelle & Orman Lutz, 5400 Hillen Drive, Oakland	4,000	160	XXVII, p. 29
Barandun							
Beaudry Placer	35	40 N.	9 W.	Madam Angele Bazet, 4 Laguna St., San Francisco.	4,000		XIV, p. 844; Bull. 92, p. 99; XXI, p. 464
Belle Josephine Placer (Slide Creek)	14	39 N.	9 W.	A. E. Westover, Callahan	6,000	160	XIV, p. 844; XXI, p. 464
Bender	18	42 N.	11 W.	7 mi. S. Etna	3,300	40	XIV, p. 826; XXI, p. 431
Better Yet				13 mi. NE. Scott Bar			XII, p. 277; XIII, p. 388
Big Bend Placer				2 mi. W. Oak Bar; H. J. Barton, James Allen, et al, Yreka	3,010		XIII, p. 388
Big Cliff	16	40 N.	10 W.	3 mi. SE. Finleys Camp	2,000	120	XXI, p. 464
Big Flat Placer	8	17 N.	7 E.	H. E. Attebery, Happy Camp	6,000	240	See text this report
Big Ledge	36	43 N.	9 W.	6 mi. NE. Greenview	1,500	20	XII, p. 277; XIII, p. 388; XIV, p. 826; XXI, p. 431
Big Rock Placer	19	18 N.	6 E.	J. V. Attebery, Happy Camp	3,000		XIV, p. 859; XXI, p. 465
Big Slide Placer		46 N.	8 W.		2,000	20	
Billups (Cleland, Brown & Billups)				11 mi. NE. Scott Bar	4,320		XII, pp. 279, 280; XIII, p. 288

Name	Sec.	Twp.	Range	Location and owner			References
Black Bear Quartz Mine		39 N.	11 W.	R. S. Phippeny, Sawyers Bar	4,200	70	VIII, pp. 620, 621; X, p. 656; XI, p. 431; XII, p. 277; XIII, p. 389; XIV, p. 826; XXI, p. 432. See text this report
Black Dike				See Blue Eagle			
Black Hawk	31	46 N.	7 W.?	5 mi. N. Gottville		40	XIV, p. 827; XXI, p. 432
Black Hawk, Hidden Treasure, Triangle				16 mi. W. Yreka, W. L. McLaughry and Fred Martin, Yreka	4,400		XXI, p. 432
Black Hawk				See Claughry			
Black Wolf Placer				Harry Walker, Clear Creek P.O.			
Blaylock & Schon Placer	12	43 N.	10 W.	Near Oak Bar	2,000	17	XXI, p. 465
Blind Lode	33, 34	11 N.	7 E.	Estate of H. J. Diggles	3,500	51	XII, p. 278; XIII, p. 390; XIV, p. 827; XXI, p. 433
Bloomer Placer				Bennett Co., Forks of Salmon	1,500		XI, p. 426; XII, p. 278; XIV, p. 390; XIV, p. 844; XXI, p. 465; VIII, p. 612
Blue Bar Placer	31	18 N.	7 E.	C. Scott Greening, Happy Camp	2,000	30	See text this report
Blue Channel Placer	19	48 N.	9 W.	Joe and Henry Slotik, Copper, Ore.	5,000	60	XXVII, p. 56
Blue Eagle	10	39 N.	10 W.	J. W. Preston, 350 Post St., San Francisco	6,500	560	See text this report
Blue Gravel Placer (Black Lead)	32	45 N.	7 W.		3,500		XIV, p. 844; XXI, p. 465
Blue Gravel Placer	35, 36, 1, 2	16 N.	7 E.	Frank Murree and John Whittaker, Happy Camp	2,000	100	See text this report
Blue Hill Placer	16	45 N.	10 W.		2,000	20	XIV, p. 845; XXI, p. 465
Blue Horn	24	45 N.	8 W.	R. D. Freshour, Yreka	4,000	40	See text this report
Blue Jay	11	47 N.	8 W.	Blue Jay Mining Co.	4,000	160	XIV, p. 827; XXI, p. 433
Blue Lead	26	42 N.	9 W.	L. H. Cory	4,000	40	XIV, p. 828
Blue Mine				See Schroeder			
Blue Nose Placer	5	13 N.	6 E.	Blue Nose Mines Co., Blue Nose P.O.			XXI, p. 465
Bonanza	14	46 N.	7 W.	Klondike Mining & Milling Co., J. P. Kleprock, Pres., Long Beach	3,000		XIV, p. 828; XXI, p. 433
Bonanza				2½ mi. E. Forks of Salmon, J. F. Wyrick and L. R. Pownell, Forks of Salmon		60	XXI, p. 433
Boot Placer	1	16 N.	7 E.	On Grizzly Creek, tributary of Indian Creek	3,750		XII, p. 278; XIII, p. 390
Boss	16	39 N.	11 W.	Mrs. T. M. Park, San Francisco	3,860	960	Address: 798 Post St.
Boulder	30	29 N.	3 W.	W. H. Wescott, Rollin	1,100	20	XXI, p. 433; see "Norcut"
Boulder Creek Placer (Fippen & Hayden)				J. M. Campbell, Dunsmuir	3,900		XXI, p. 434
Boulder Point Placer				3 mi. S. Callahan, Fippen & Hayden, Callahan	5,000	60	XX, p. 180; XXI, p. 465
Bowersox Placer	2	11 N.	6 E.	Joseph Weeks, Scott Bar	600		XX, p. 466
Bowser	11	47 N.	8 W.	W. P. Bowersox, Somes Bar	5,000	40	XIV, p. 845; XXI, p. 466
Boyle	8	45 N.	8 W.	C. L. Bowser, Hilt	5,000	40	See text this report
Brazil Placer	11	45 N.	7 W.	See Roff Mine	3,000	80	X, p. 656; XI, pp. 444, 445; XII, p. 278; XIII, p. 390; XIV, p. 828; XXI, pp. 434, 453
Bridger				4 mi. from mouth McKenny Creek			VIII, p. 612; XIII, p. 391; XIV, p. 859; XXI, p. 466
Brown & Billhaps				14 mi. SE. Sawyers Bar, T. H. George, Cecilville			
Brown & George Placer				See Cleaver; also Golden & Eveleth			XII, p. 279
Brown Bear	3	18 N.	6 E.?	H. E. Attebery, et al., Happy Camp			XII, p. 391; XIV, p. 845; XXI, p. 466
Bull Dog Placer				See Siskiyou County Mine			
Bumblebee Placer				On Bumblebee Creek			XII, p. 279
Bumblebee				See Joubert Placer			
Burns Placer							
Burton	1	44 N.	6 W.?	R. H. Burton, Fort Jones			XXI, p. 434

REPORT OF STATE MINERALOGIST

TABLE OF GOLD MINES AND PROSPECTS. SISKIYOU COUNTY — Continued
With References to State Mineralogist's Reports

Name of mine	Sec.	Twp.	Range	Remarks on location or owner	Elevation, feet	Area, acres	References
Buzzard Hill	4, 5	15 N.	7 E.	Buzzard Hill Mine, Inc., P. M. Tolman, Supt.	1,500	240	XXI, p. 434. See text this report
	18	46 N.	6 W.	Happy Camp	2,290	20	XIII, p. 291; XIV, p. 861
California Bar Placer	16, 17	39 N.	11 W.	Vene Gold Bar Mining Co.			XII, p. 282; XIII, p. 402; XIV, p. 828
California Consolidated (Golden Bull)				California Consolidated Min. Co., W. H. Young, Pres., Oakland	3,200	384	XII, p. 279; XIII, p. 392; XIV, p. 845; XXI, p. 466
CalOro Dredging Co.	11, 13	43 N.	10 W.	See Gardella Dredge	3,000	522	XII, p. 279; XIII, p. 392
Campbell Placer				H. A. Weed and others		115	XIV, p. 829; XXI, p. 435
Cannon				6½ mi. SE. Forks of Salmon	5,000		VIII, p. 614; XIII, p. 392
Cape Cod	18, 19	45 N.	7 W.	23 mi. W. Trinity Center	5,150		XII, p. 279; XIII, p. 392
Carino							
Carl & Shaw				5 mi. SE. Sawyers Bar			
Carpenter & Robertson Placer				Near Cecilville, S. Carpenter & E. M. Robertson, Cecilville			XIV, p. 845; XXI, p. 466
Cartwright & Phillips (Mountain Belle)	2	47 N.	5 W.	16 mi. NE. Yreka	3,500		XI, p. 445; XIII, p. 392
Cassaday				J. L. Corbett, Hilt	5,500		XVII, p. 533; XXI, p. 435
C. B. & Q.				½ mi. S. Oro Fino			XIII, p. 393
Cecilville Placer (Sightman)				Near Cecilville	2,500	40	XIV, p. 845; XXI, p. 466
Central	35	48 N.	8 W.	See Union	4,000	140	XI, p. 829; XXI, p. 435; XXVII, p. 30. See text this report
Challenge		47 N.	8 W.	C. R. Wiegel, Redding			XIII, p. 393
Champion	33	12 N.	6 E.	2 mi. NW. Deadwood	3,720	180	XIV, p. 829; XVII, p. 533; XXI, p. 435
Chapman	30	40 N.	7 W.	W. F. & J. A. Hunter, Orleans; J. & H. H. Chapman, Hornbrook, and T. L. Petersen			XX, p. 180; XXI, p. 436
Charter Oak	27	45 N.	8 W.	10 mi. SW. Yreka	5,000	240	XIII, p. 393
Cherry Hill				G. A. Reichman, Fort Jones	4,700		XXI, p. 436; XXVII, p. 31; XIV, p. 829. See text this report
China Creek Placer (Reeve)	5, 6, 7, 8	16 N.	8 E.	C. E. Reagan, Happy Camp	4,100	40	XIII, p. 393; XIV, p. 845
China Point Placer	36	16 N.	8 E.	C. Scott Greening, Happy Camp	1,300	195	See text this report
Classic Hill		18 N.	6 E.		1,500	310	VIII, p. 599; XI, p. 443; XIII, p. 394; XIV, p. 846; XXI, p. 466. see text this report
Claughry				14 mi. NW. Yreka	2,100		XIII, p. 394
Cleaver (Brown Bear or Golden & Eveleth)	14	39 N.	11 W.	3 mi. above mouth McKenny Creek	4,000	140	XI, p. 433; XII, pp. 279, 283; XIII, p. 402; XXI, p. 435
Cleland					4,800		XII, p. 280
Clyburn	8	46 N.	7 W.	S. F. Clyburn, Klamath River	2,000	70	See text this report
Collateral				1 mi. W. Yreka			VIII, p. 630
Collins Ranch Placer (Ambrose Min. Patent)	6, 7	16 N.	8 E.	W. E. Collins, Scad	1,100	89	See text this report

Name	Sec.	Twp. N.	Range	Owner/Operator	Elev.	Acres	References
Columbia— (including Good Enough and Insurance)	1	45 N.	10 W.	Adrian J. Fisher	5,500	23	XII, p. 280; XIII, p. 394; XXI, p. 437; XIV, p. 829
Commodore	26, 27, 34, 35	46 N.	9 W.	H. J. Barton, Yreka	3,500	220	XIII, pp. 394, 403, 408; XXI, p. 437; XXVII, p. 31. See text this report
Condensed	12	38 N.	11 W.	W. H. Cady	2,400	60	XIV, p. 830
Connor	18	43 N.	9 W.	J. Connor	3,000	20	XIII, p. 395; XIV, p. 830
Consolidated Seiwash	6	44 N.	8 W.	H. Madison	3,200	20	XIII, p. 395; XIV, p. 861
Contact & Bonanza	19	46 N.	6 W.	B. G. Reedie, Yreka	2,100	90	XXI, p. 437; XXI, p. 467
Conzetti Placer				In Cecilville District, G. A. Conzetti, Cecilville			
Cook, Abel W.				See under "A"			
Corbett	2	47 N.	8 W.	J. L. Corbett, Hilt	5,400		XII, p. 280; XIII, p. 396; XXI, p. 437; XXVII, p. 32
Corey Bros.	29	40 N.	7 W.	Corey Bros., c/o George H. Corey, Callahan	5,000	80	
Corey	21	40 N.	7 W.	Abe Goldberg, Etna			
Corey	30, 24	40 N.	8 W.	Corey Bros., c/o Geo. H. Corey, Callahan	5,000		XXI, p. 438
Cornish		10 N.	7 E.	On Big Humbug Creek			See text this report
Cory				See Roff			
Cosmos Mines Development Co.				See Hansen			
Crajo Placer	11	10 N.	7 E.	Bennett Co., Forks of Salmon	1,200	39	VIII, p. 612; XI, p. 427; XII, pp. 280, 282; XIII, p. 396; XIV, p. 846; XXI, p. 467
Crary				See S. R. Crary under "S"			
Crawfish Gulch	18	45 N.	10 W.	Sam Bratt, Oak Bar		40	XVII, p. 533; XXI, p. 438
Crawley	22	40 N.	8 W.		3,500		VIII, p. 629; XIII, p. 396; XIV, p. 830; XXI, p. 438
Croesus Placer (Hickey)				6½ mi. W. Sawyers Bar, John Teukert, Sawyers Bar		20	XXI, p. 467
Cronin Placer				See Gallia Placer			
Crumpton	2	16 N.	7 E.	See "Blue Gravel" in text		200	
Cub Bear and Blue Jeans	9	40 N.	10 W.	Siskiyou Syndicate, I. J. Luce, Pres., 618 2d Ave., Seattle, Washington	5,000	640	XXI, p. 438; XIV, p. 830
Cummings				See Oro Grande			
David Ledge				13 mi. NE. Scott Bar	3,100		XIII, p. 396
Davis Consolidated Mines (Placer)	2, 9, 10, 11, 15, 16	16 N.	7 E.	Estate of Reeves Davis, c/o W. F. Davis, 427 J St., Sacramento	1,200	603	XIV, p. 846, XXI, p. 467. See text this report
Deep Channel Placer (Taylor & Maplesden)	36	46 N.	11 W.	Deep Channel Mining Co., J. L. McKittrick, Pres., Portland, Oregon	1,500	40	XII, p. 283; XIII, p. 429; XIV, p. 861
Deer Lodge				6 mi. S. Sawyers Bar	4,050		XIII, p. 396
Del Norte				See Richardson			
Denny Placer	17, 20, 21, 29	40 N.	8 W.	Scott River Dredge Co.	3,200	400	XIV, p. 846
Dewey	6	41 N.	6 W.	F. A. Wright, 1744 Franklin St., Oakland	6,800	100	XIV, p. 831; XXI, p. 438; XXVII, p. 32. See text this report
Dick Morris Placer (Joe Ramus Placer)	35	46 N.	11 W.	J. B. Nowdesha and Mrs. S. D. Johnson, Hamburg Bar	1,500	59	XIV, p. 848; XXI, p. 469
Doggett Placer				Near Oak Bar	2,000		XXI, p. 469
Dollarite	5	17 N.	8 E.	G. H. Lindley, Seiad	4,000	80	
Doolittle Placer	26	46 N.	9 W.	In Happy Camp District, M. Doolittle		40	
Double Eagle and Little Quartz				H. J. Barton, Yreka	3,000		XIII, p. 397; XIV, p. 831; XXI, p. 435
Dundee				14 mi. NW. Yreka	4,000		XIII, p. 398
Dunnigan Placer	20	12 N.	6 E.	36 mi. S. Happy Camp		20	XIV, p. 848; XXI, p. 469
Easter				In Humbug District, Frank Zollihofer, Yreka			XXI, p. 438
East Fork Ledge				5 mi. E. Callahan			XIII, p. 398

TABLE OF GOLD MINES AND PROSPECTS, SISKIYOU COUNTY Continued

With References to State Mineralogist's Reports

Name of mine	Sec.	Twp.	Range	Remarks on location or owner	Elevation, feet	Area, acres	References
Eastlick Placer	7, 8	43 N.	9 W.	Reichman Mercantile Co., Fort Jones	3,000		VIII, p. 608; XIII, p. 398; XIV, p. 848; XXI, p. 469
Eli	4, 5, 8, 9	45 N.	8 W.	4½ mi. W. Yreka			XIII, p. 398
Eliza	9			R. H. DeWitt, Yreka	4,000		XIV, p. 831; XIX, p. 138; XXI, p. 439; XXVII, p. 32. See text this report
Elk Creek	3	45 N.	7 W.	Elk Creek Mining Co., J. E. Harmon, Secretary, Yreka			XIV, p. 831
Elk Creek Placer	15	16 N.	7 E.	Address, Roseburg, Oregon	3,000	100	XII, p. 281; XIII, p. 398; XIV, p. 861
Elliott Creek Mines Co.	18	48 N.	10 W.	1½ mi. W. Sawyers Bar	1,100	60	See text this report
Elliott Placer (Golden Nugget)	33	11 N.	7 E.	A. E. Ellston	3,900	450	XIV, p. 848; XXI, p. 469
Ellston Placer				Near Yreka	2,200	20	XIV, p. 848
El Oro No. 3 Dredge					1,200	30	Bull. 85, pp. 36, 37; XXI, p. 490
Empire Bar Placer	27, 34	46 N.	8 W.	J. W. Wright, Etna	3,000	20	VIII, p. 559; XIV, p. 861
Empire		41 N.	9 W.	2 mi. N. Gottville	1,900	60	See text this report
Empire				2 mi. NE. Sawyers Bar, Geo. F. Townes and W. E. Tebbe, Weed	3,500		XII, p. 281; XIII, p. 399
Enterprise				Hollis Anderson, Scott Bar	2,500		
Enterprise Placer	21	45 N.	10 W.	See Queen of Sheba	3,900	100	XXI, p. 439
Espey Placer	25	47 N.	9 W.	Espey Mining Co., Seattle, Washington	2,500	100	VIII, p. 623; see text
Eton				6 mi. SW. Sawyers Bar, Dunphy Estate, Sawyers Bar and W. J. Durch, Rollin		50	XIV, p. 849
Eureka and W. J. D. (Firebug)				5½ mi. NE. Happy Camp, C. A. Evans, Marshall Crawford			XXI, p. 439
Evans Placer (Berry)				See Union		20	XIV, p. 862
Evening Star	26	46 N.	11 W.	J. H. Everill	1,500	40	XII, p. 281; XIII, p. 400; XIV, p. 862
Everill Placer				See Humpback			XXI, p. 439
Fagundez				Long Gulch near Hawkinsville	4,000		XI, p. 446; XII, p. 282
Fairchild				Divide between Empire and Hungry Creeks	3,400		XIII, p. 400; XXI, p. 439
Fairy Queen				11 mi. NW. Yreka			
Falcon	22, 23	48 N.	10 W.	S. K. Hine and L. B. Miller, 924 National City Bank Bldg., Cleveland, Ohio	4,500	210	XXVII, p. 57
Fawcett Mines Co. (Pennsylvania Placer)				See Schroeder			
Fidelity Metals Corp.				See Boulder Creek Placer	1,500	20	XIV, p. 849; XXI, p. 470
Fippen & Hayden				1 mi. W. Forks of Salmon, Bennett Co., Forks of Salmon			XIII, p. 401; XVIII, p. 297; XXI, p. 439
Fir Tree Placer				On Sucker Creek in Humbug District	2,700		See text this report
Flag	34	45 N.	7 W.				XI, p. 446; XIV, p. 832; XXI, p. 439
Fledderman Bros	5	45 N.	8 W.	W. of Humbug			
Fleetwood and Nannie S. (Jackson)							

Name	Sec.	Twp. N.	Range	Location / Operator			References
Foster	7, 8, 9, 16, 17, 18	10 N.	8 E.				
Forks of Salmon River (Placer)				See Salmon River Mines Co.			
Forks Placer	31	47 N.	12 W.	Forks of Salmon River Mining Co.	1,300	520	XIV, p. 850
Fort Goff Placer	16	44 N.	9 W.	At Forks of Salmon, Bennett Co.	1,200	51	XII, p. 287; XIII, p. 401; XIV, p. 849; XXI, p. 470
Franklin				14 mi. W. Hamburg Bar, G. Martin	3,000	80	VIII, p. 506; XII, p. 283; XIII, p. 401; XIV, p. 850
Franks & Moncton						40	XIV, p. 832; XXI, p. 440
Frazier Placer				See Osceola			
French Bar Placer	17	45 N.	10 W.	See Woodfill & Barry	1,700	20	XIV, p. 862
French Bar Placer	24	15 N.	7 E.	Z. E. Russell	2,000	20	
Galena				Harry D. Maltis, Happy Camp	4,000		XIII, p. 401; XXI, p. 440
Gailla Placer (Paddy Cronin)				12 mi. NW. Yreka		120	
Gardella Dredge				45½ mi. W. Sawyers Bar, A. Jacquemart. 320 Market St., San Francisco.	1,900		XXI, p. 471; XXVII, p. 56. See text this report
Gardiner & Deming (Placer and Quartz)	7, 18	43 N.	9 W.	Near Yreka			
Gearhart Placer				On Horse Creek, 4 mi. above mouth, H. J. Barton, Yreka	3,000	150	XIII, p. 402; XIV, p. 850
Geeshan & Kellner Placer	30	40 N.	11 W.	Joe Davidson and J. J. Skehan, Gottville	1,850	120	XIII, p. 402; XXI, p. 472; XIV, p. 850; XXI, p. 473
Gibson Bar Placer	6	46 N.	7 W.	4 mi. SW. Oak Bar, J. D. Nowdesla, Hamburg	2,000	20	XXI, p. 473
Gilfeather				G. A. Dannenbrink et al., Etna	2,000	20	XXI, p. 440; XXVII, p. 34
Gilta (Gold Hill)	12	9 N.	7 E.		3,000	60	XXI, p. 440; XXVII, p. 34; XI, p. 622; XIII, p. 429; XIII, p. 407 (Hungry Hill); XIV, p. 833; XXI, p. 440. See text this report
Gold and Nickel	35	42 N.	7 W.	James F. Furlong, Fred Gould et al., Gazelle	3,000	200	XXVII, p. 34
Gold Ball Mining Co.	2	41 N.	7 W.	F. A. Gowing, Manager, Sawyers Bar	5,000	80	X, p. 657; XI, p. 432; XII, p. 282. See text this report
Gold Bank Placer		39 N.	11 W.	See Woodfill & Barry	4,000		
Gold Bug				See Lanky Bob			
Gold Bug	6	46 N.	6 W.	C. V. Clark, c/o Mrs. V. Barbour, Multnomah, Oregon.	3,663	60	XXVII, p. 35
Golden Age				E. W. Cooper and F. M. Kirkland, Walker P.O.		60	
Golden & Eveleth	4	45 N.	9 W.	See Cleaver			
Golden Ball				See California Consolidated			
Golden Eagle Claim				See Golden Age			
Golden Eagle (Indian Creek)	11	44 N.	9 W.	Geo. A. Milne, Fort Jones	3,000	60	VIII, p. 625; XII, p. 282; XIV, p. 832; XXI, p. 440; XXVII, p. 35. See text this report
Golden Rule	10	47 N.	8 W.	R. G. Harrison, Klamath River	4,000	40	XXI, p. 474
Golden Rule and Scott Bar Placer				At Scott Bar, Turner Bros., Yreka			XIV, p. 832; XXI, p. 441
Golden Seal	18	43 N.	9 W.	See Gilta	3,000		XIV, p. 832; XXI, p. 441
Golden West	15	39 N.	10 W.	Bigelow Bros., Sawyers Bar	7,000		
Gold Hill				M. Andrews			XIV, p. 850; XXI, p. 473
Gold Hill Placer	26, 29	40 N.	11 W.	J. M. Weaver, Yreka	2,400	20	XIV, p. 862
Gold Lead Placer	8	45 N.	10 W.	Chas. Dillstrom, Yreka	1,600	80	See text this report
Gold Leaf	26	46 N.	7 W.?	Kenneth K. Ash, Yreka		20	XXI, p. 473
Gold Leaf Placer		45 N.	7 W.	10 mi. S. Forks of Salmon	4,000		XXVII, p. 40
Gold Road	23	45 N.	8 W.	See Mountain Laurel	2,000	110	X, p. 657; XI, pp. 429, 431; XII, pp. 283, 285; XXI, p. 441
Gold Run (Hungry Hill)				See Commodore; also Spencer			
Gold Standard, Inc.				S. of Gilta			See text this report
Goodenough							
Good Hope							

TABLE OF GOLD MINES AND PROSPECTS, SISKIYOU COUNTY - Continued

With References to State Mineralogist's Reports

Name of mine	Sec.	Twp.	Range	Remarks on location or owner	Elevation, feet	Area, acres	References
Goodman				On Indian Creek near Classic Hill, Joe Goodman, Happy Camp	2,500		XXI, p. 441
Gordon Placer	17	18 N.	8 E.	5 mi. NE. Happy Camp, C. Gordon		60	XII, p. 283; XIII, p. 403; XIV, p. 350
Grattan Mine				12 mi. S. Sawyers Bar			XII, p. 283
Gravel Mines, Ltd.	5, 8, 9	16 N.	8 E.	J. C. Moulton, Supt., Seiad	1,300	100	See text this report
Great Northern (Extension of Siskiyou)				6½ mi. NW. Yreka	2,900		XIII, p. 403
Green Mountain				2 mi. SW. Happy Camp	3,220		XIII, p. 403
Green Mountain				In Quartz Valley District			XI, p. 437
Green's Bar Placer				Steve Green, Happy Camp	1,000	40	VIII, p. 600; XIII, p. 403; XIV, p. 851; XXI, p. 474
Grider Placer	10, 11	16 N.	7 E.	J. B. Grider	1,400	245	XIV, p. 833; XXI, p. 441
Grizaly Gulch	14	46 N.	12 W.		7,000		
Grubstake Placer	16	44 N.	9 W.	C. H. Attebery, Happy Camp	1,500	20	XIII, p. 404; XIV, p. 833; XXI, p. 441; XXVII, p. 41. See text this report
Gumboot	16	45 N.	9 W.	Jack McInnes and Philip McCool, Scott Bar	4,000	100	
Guy Ford				See Jumbo			
Hadlof				13 mi. NE. Scott Bar	3,150		XIII, p. 404
Haley Placer (Halstead)	7	16 N.	7 E.?	10 mi. SW. Happy Camp, M. Doolittle		40	VIII, p. 601; XIII, p. 404; XIV, p. 851
Hammer Placer	36	46 N.	11 W.	Johnson et al., Hamburg Bar	1,500		XIV, p. 851; XXI, p. 474
Hansen	1, 12	9 N.	7 E.	Roberts & Hagland	2,500	60	VIII, p. 622; X, p. 657; XI, pp. 429-431; XXI, p. 442. See text this report; XIV, p. 833
Happy Home				See Maplesden			
Hard Luck Placer	9	17 N.	7 E.	H. Miller, Happy Camp	1,500	20	XIII, p. 404; XIV, p. 851
Hardscrabble Placer	1, 12	44 N.	9 W.	J. D. Duane	3,000	20	
Hardscrabble	16	40 N.	10 W.	A. M. Bailey, E. J. Miley, Sr., Los Angeles; and John Nefroney, Etna	5,200	100	XXI, p. 442; XIV, p. 833
Hartstan	11	40 N.	9 W.	7 mi. SW. Etna			XIII, p. 405
Hathaway	36	47 N.	7 W.	Fred Salter, 317 S. Olive St., Los Angeles	3,500		XXVII, p. 42. See text this report
Hazel				Hazel Gold Mining Co., Rm. 456, 79 New Montgomery St., San Francisco	3,500	80	XIV, p. 833; XXI, p. 442; XXVII, p. 42. See text this report
Heeler				See Oregon			
Herndon Placer	17	46 N.	7 W.	A. C. Herndon, Hornbrook	2,000		XXVII, p. 58. See text this report
Hibernia				2½ mi. SE. Sawyers Bar			XII, p. 284
Hickey	24, 25	40 N.		8 mi. SE. Sawyers Bar, Ed. Hickey, Sawyers Bar	4,060	120	XXI, p. 443. See text this report
Hickey Placer			12 W.	Pike & Hickey	2,000	20	XIII, p. 405; XIV, p. 852
Hickox Hydraulic				15 mi. S. Cottage Grove			XIII, p. 405
Hicks	2	46 N.	7 W.	Foster Bros.	6,500	60	XIV, p. 834; XXI, p. 443
Highland (Old Highland)	12	39 N.	10 W.				XIV, p. 834; XXI, p. 443
Highland	9	40 N.	10 W.	Highland Mining Corp., 904 Garfield Bldg., Los Angeles	6,000	320	XXI, p. 443

Name	Sec.	T.	R.	Owner and location	Elev.	Acres	References
Highland	25	46 N.	7 W.	½ mi. NW. Deadwood, G. A. Reichman and W. H. Young, Fort Jones			XIV, p. 834; XXI, p. 443
Hiyou (Davis)	30	45 N.	8 W.	H. Young, Fort Jones		40	XIII, p. 405. See text this report
Hiyou Placer				8 mi. NW. Fort Jones, G. A. Reichman and W. H. Young, Fort Jones		40	XIII, p. 405; XIV, p. 852. See text this report
Hoboken	21	45 N.	8 W.	Estate of H. J. Diggles	4,500		XII, p. 284; XIII, p. 406; XIV, p. 835; XVIII, p. 496; XXI, p. 444
Hogan				16 mi. SW. Etna	4,400		XXI, p. 445
Homestake	21	41 N.	10 W.	16 mi. SW. Etna, R. S. Taylor, Yreka	4,400		XXI, p. 444; XIV, p. 835
Hooper Hill (Preckel)		45 N.	10 W.	H. Preckel	1,800	20	XIV, p. 852
Hoozier Hill Placer (Willard Drift)	36	46 N.	11 W.	J. B. Nowdesha	1,500	20	XII, p. 294; XIII, p. 432; XIV, pp. 852, 857; XXI, p. 474
Huey Hill Placer	25	18 N.	6 E.	D. Huey	2,000	40	XIII, p. 407; XIV, p. 852. See text this report
Humpback (Fagundez)	8	39 N.	11 W.	Alice Fagundez	4,000	80	VIII, p. 619; XI, p. 432; XII, p. 281; XIII, p. 386; XIV, p. 835; XXI, p. 445
Hungry Hill (Gold Run)				8 mi. S. Forks of Salmon	3,350		XI, pp. 429, 430; XII, pp. 283, 285; XIII, p. 407
Hunter & Downey				5 mi. S. Sawyers Bar			XII, p. 285; XIII, p. 407
Hunter's Paradise				See Paradise			
Ida May				4 mi. S. Sawyers Bar, A. J. Ball, Rollin	3,700	54	XXI, p. 445. See Norcal Mining Co., also
Imperial Heights Placer	32	15 N.	7 E.	2 mi. W. Sawyers Bar, Ed. Hickey, Sawyers Bar	2,100	40	XII, p. 284; XIV, p. 852; XXI, p. 474
Independence				Buzzard Hill Mine, Inc., P. M. Tolman, Supt., Happy Camp			
Indian Girl	14	46 N.	7 W.		1,225	140	XXI, p. 446. See text this report
Insurance				See Commodore	2,300	40	See text this report
Inyo	2	45 N.	8 W.		3,000		XIV, p. 835; XXI, p. 447
Iron Dike	22	48 N.	8 W.	D. M. Watt, Phoenix, Oregon	4,000	80	See text this report
Ironsides	26	45 N.	8 W.		4,500		XII, p. 285; XIII, p. 408; XIV, p. 835; XXI, p. 447
Jack Lowden Placer				See Lowden			See
Jehova	5	16 N.	8 E.	9 mi. NW. Yreka	3,000		XIII, p. 408
Jim's Seattle Placer	6	45 N.	10 W.	J. A. King, Happy Camp	1,100	20	See text this report
Joe Ramus	12	43 N.	10 W.	G. A. Milne, Fort Jones	1,500	60	XIV, p. 853
Johnson & China Paul	18, 19	43 N.	9 W.		3,000		VIII, p. 826; XIV, p. 835; XXI, p. 447
Johnson & Lewis		43 N.	11 W.	John J. Johnson and others, Etna	3,500	120	VIII, p. 626; XII, p. 285; XIII, p. 408; XXVII, p. 43. See text this report
Joubert Placer (Peterson)	33	40 N.	11 W.	3 mi. S. Sawyers Bar, L. J. Joubert, Sawyers Bar	2,500	87	XXI, p. 474
Julia	1	39 N.	11 W. ?	See Underland			XII, p. 285; XIII, p. 409; XIV, p. 840; XXI, p. 447. See text this report
Jumbo				Guy Ford, Weed	2,900		See text this report
Kanaka Hill Hydraulic	27, 33	16 N.	7 E.	Steve S. Green, Happy Camp	1,500	100	See text this report
Kangaroo	29	40 N.	7 W.		5,500		XIV, p. 836; XXI, p. 448
Katie May	24	45 N.	8 W.		4,000	20	XIV, p. 409; XIV, p. 836; XXI, p. 448. See text this report
K. C.				See Golden Age			
Keaton	24	15 N.	7 E.	9 mi. E. Sawyers Bar, Harry D. Maltis, Happy Camp	4,400	100	See text this report
Keno Placer	30	46 N.	6 W.	C. E. Weston, Yreka, Chas. G. Peterson, 1825 Oak St., San Francisco, Chas. Magill, Redding	2,000	20	XXI, p. 448
Keynote				9 mi. SW. Edgewood	3,000		XIII, p. 410
Kiernan	7	17 N.	7 E.	Harry D. Maltis, Happy Camp	6,000	120	See text this report
King Jade	14	38 N.	12 W.	King Solomon Mines Co., Roy N. Bishop, Pres., 411 Crocker Bldg., San Francisco, or Black Bear P. O.	2,000	120	XXI, p. 448
King Solomon					3,500	800	XIV, p. 836; XXI, p. 448. See text this report

TABLE OF GOLD MINES AND PROSPECTS, SISKIYOU COUNTY—Continued

With References to State Mineralogist's Reports

Name of mine	Sec.	Twp.	Range	Remarks on location or owner	Elevation, feet	Area, acres	References
King Solomon	32	41 N.	7 W.	12 mi. NW. Yreka	4,000		XIII, p. 410
King Tut				T. A. Lloyd, Gazelle	5,000		See text this report
Kinkaid				9 mi. SW. Callahan, John Kinkaid, Sawyers Bar	4,000	220	XXI, p. 448
Klamath				1½ mi. SW. Rollin, A. J. Ball, Rollin. See "Norcal" also.		20	
Klamath Placer Co.	16	46 N.	7 W.		4,500	80	VIII, p. 620; XI, p. 432; XIII, p. 410; XXI, p. 448
Klamath River Gold Mining Co.	15	46 N.	7 W.	Klamath River Gold Mining Co.	2,000	68	See text this report
Klein Placer (Casey)	4	39 N.	11 W.	Woodfil & Luddy	2,000	20	XIV, p. 862
Klondike Mining & Milling Co.				See Bonanza	3,000	50	XII, p. 286; XIII, p. 411; XIV, p. 853
Knownothing	15	16 N.	7 E.	8 mi. SE. Forks of Salmon	3,075		VIII, p. 622; XI, p. 431; XII, p. 286; XIII, p. 411
Knownothing Placer				C. E. Reagan, Happy Camp	1,000		See text this report
Lango				16 mi. W. Yreka			XII, p. 286
Lanky Bob				5 mi. SE. Sawyers Bar, G. T. Salsbury, Etna	3,300		XXI, p. 449. See text this report
Last Chance				4 mi. W. Oro Fino			VIII, p. 624; XII, p. 286; XIII, p. 411
Last Chance (Grattan)				6 mi. E. Forks of Salmon			XIII, p. 283; XIII, p. 411
Last Chance (Stewart)				7½ mi. N. Fort Jones	3,225		XIII, p. 411
Lewis				¾ mi. N. Oro Fino			XIII, p. 412
Liberty				4 mi. S. Sawyers Bar	3,400		XII, p. 286; XIII, p. 412
Lima				4 mi. W. Yreka	2,400		XIII, p. 412
Lincoln, Abe, Placer				See under "A"			
Little Bonsa	9	45 N.	7 W.	C. W. Gordon, Yreka	3,600	30	XXI, p. 449; XIV, p. 836
Little Crumpton Placer	1	16 N.	7 E.	Clifford & Leonard Crumpton, Happy Camp	1,200		See text this report
Little Queen (Erno)				Mouth Evans Gulch in Quartz Valley			XII, p. 281; XIII, p. 413
Live Yankee				5 mi. S. Sawyers Bar			XI, p. 432
Livingston & Everton				In Happy Camp District, T. E. Ahlstrom, Ashland, Oregon			
Long Tom Placer	11, 12, 14	40 N.	11 W.	Judin Johnson, Sawyers Bar	2,400	140	XXI, p. 450
Lowden Placer		46 N.	12 W.	Ariel Lowden, Seiad	1,400	393	XXI, p. 476
Lowden Placer	35	45 N.	11 W.	J. F. Lowden, Hamburg Bar	1,800	60	XXI, p. 476. See text this report
Lowden, Jack, Placer	13	46 N.	12 W.	S. R. Huey, Newcastle, Pa.	1,500	73	XIV, p. 852. See text also
Lucky Bob Placer	2	46 N.	7 W.	Joseph Freshour, Gottville	1,900	20	XII, p. 287; XIII, p. 413; XIV, p. 862; XXI, p. 477
Lucky John				On N. Fork Humbug Creek	3,000		XIII, p. 413
Lucky Strike	33	41 N.	10 W.	Lucky Strike Mining Co., T. Eagerly, Pres., Los Angeles	5,000	80	XIV, p. 836
Lucky Strike				See Osgood			
Lumprey	22?	47 N.	8 W.	Calstrom & Stelik, Portland, Ore.	3,500		XII, p. 287; XIII, p. 414
Mabel				7 mi. NE. Scott Bar	4,875		XIII, p. 414
Magnitude				3 mi. W. Yreka	2,900		
Malloy (Oregonian)				3½ mi. NE. Sawyers Bar, Dan Malloy, Sawyers Bar	5,000		XII, p. 289; XIII, p. 416; XIV, p. 838; XXI, p. 450

Name	Sec.	T.	R.	Owner or location			References
Maun & Ross Placer (Reynolds Creek)	21	12 N.	6 E.	Ed. Mann & Nelson Ross, Orleans	------	100	XXI, p. 477
Maplesden Placer (Happy Home)	31	46 N.	10 W.	Maplesden Bros	1,500	40	VIII, p. 594; XIV, p. 851; XXI, p. 477
Marrian & Goodale	15	40 N.	10 W.	------	5,500	80	XIV, p. 837
Mary Alma Placer	10	46 N.	12 W.	Carl T. Frey, Seiad	1,400	20	XXI, p. 477
Masonic Bar Placer	4	46 N.	12 W.	Walter M. Creed, Seiad	1,300	108	XXI, p. 477
Mat-a-pan	17?	45 N.	10 W.	M. O. Payne, Scott Bar	------	20	See text this report
Mattoon Placer	34	45 N.	9 W.	R. A. Mattoon, Fort Jones	3,600	60	XXVII, p. 60
Mayland Mining Co.				See White Bear			
McCann Placer				9 mi. S. Happy Camp, F. B. McCann, Happy Camp	1,000	20	XXI, p. 479
McClaughry	8	45 N.	8 W.		5,000		XIV, p. 837; XXI, p. 450
McConnell Bar				See Klamath Placer Min. Co.			
McCook Placer	1	45 N.	8 W.	Charles Benbeck, Yreka	2,660	24	XXI, p. 479
McGuffey Placer	16	45 N.	10 W.	T. G. McGuffey	1,700	40	XII, p. 287; XIII, p. 415; XIV, p. 853
McKeen				See Oro Grande			
McKenzie				5½ mi. N. Gottville, W. H. Herndon			XXI, p. 450
McMahon Placer	25	40 N.	9 W.	C. A. Crowley	3,500	40	XII, p. 287; XIII, p. 415; XIV, p. 863
Metropolitan				4½ mi. SE. Sawyers Bar			XII, p. 287; XIII, p. 415
Michigan-Salmon Hydraulic (Red Hill)				2 mi. SE. Forks of Salmon, Bennett Co., Forks of Salmon			XIV, p. 853; XXI, p. 479
Might, Wm.	2	10 N.	7 E.	Near Classic Hill	1,300	600	XXI, p. 450
Milich Placer	34	44 N.	9? W.	P. Milich	2,000	20	XII, p. 288; XIII, p. 415; XIV, p. 854
Miller Placer				Jackson & Biedenbeck	1,200	20	XIII, p. 415; XIV, p. 863
Miller				⅜ mi. NW. Oro Fino			VIII, p. 624; XIII, p. 415
Minerals Recovery Corporation, Ltd.				See text this report			
Mint	8	39 N.	11 W.	1½ mi. W. Yreka	3,000		XIII, p. 416
Monarch				G. R. Godfrey	3,300	40	XIV, p. 837
Monarch				15 mi. SE. Forks of Salmon, L. W. Godfrey, Sawyers Bar			XXI, p. 450
Montezuma Placer	20	40 N.	8 W.	E. W. Morgan, Oak Bar	3,500	20	VIII, p. 612; XII, p. 288; XIII, p. 416; XIV, p. 863
Morganza Placer	13	46 N.	10 W.	13 mi. SE. Scott Bar	3,400	60	XXI, p. 490
Morning Star					3,270		XIII, p. 416
Morrison & Carlock (Little Queen)	13	43 N.	10 W.	Geo. A. Milne, Fort Jones	2,800	60	XIII, p. 413; XIV, p. 837; XXI, p. 450; XXVII, p. 44. See text this report
Mosher & Lowell				On Horse Creek			XXI, p. 450
Mountain Belle				1 mi. up from junction of Eliza Fork with N. Fork Humbug Creek			XI, p. 445; XII, p. 288
Mountain King				12 mi. N. Orleans, D. C. Weaver, Orleans	4,000		XXI, p. 452
Mountain Laurel		39 N.	11 W.	Gold Standard, Inc., 2434 1st Ave, So., Seattle, Washington			VIII, p. 619; XI, p. 432; XXI, p. 451
Mountain Lily				In Salmon River District, G. A. Conzetti, Cecilville	3,800		XVII, p. 534; XXI, p. 451
Mount Bolivar				5 mi. SW. Callahan	6,000	140	XIII, p. 416
Mount Vernon	26	45 N.	8 W.	Kathryn J. Pfeiffer, c/o K. K. Ash, Yreka	4,620		XXI, p. 452; XXVII, p. 45; XIV, p. 837. See text this report
Muck-a-Muck Placer	18	16 N.	8 E.	4½ mi. E. Happy Camp, Grant Smith, 693 Mills Bldg, San Francisco	1,430	110	VIII, p. 598; XII, p. 288; XIII, p. 416

TABLE OF GOLD MINES AND PROSPECTS, SISKIYOU COUNTY — Continued

With References to State Mineralogist's Reports

Name of mine	Sec.	Twp.	Range	Remarks on location or owner	Elevation, feet	Area, acres	References
Mullen	2	40 N.	9 W.	M. P. Mullen, Gazelle, and Otto Schmale, Chico	3,000	60	XXVII, p. 47
Multum in Parvo Placer				5½ mi. W. Sawyers Bar, Wm. Wike & Samuel Mathis, Sawyers Bar	2,000	60	XXI, p. 480; VIII, p. 622
Mystery	13	10 N.	7 E.	Near Black Bear	1,300	20	XIV, p. 854; XXI, p. 480
Native Son Placer				M. L. Mills, Forks of Salmon	4,300		XIII, p. 417
Nebraska	7	46 N.	6 W.	10 mi. SW. Yreka	3,000	122	See text this report
Nebraska	10	44 N.	9 W.	Geo. H. Dern, R. S. Stryker, Yreka	7,100	40	XIV, p. 826
Neil, Ben				B. Neil, Fort Jones			XXI, p. 452
Neilon & Putnam				4½ mi. SE. Sawyers Bar, Miles Neilon & W. Putnam, Sawyers Bar	3,500		XI, p. 446; XIII, p. 417
Nelson (Del Monte)	19	15 N.	7 E.	12 mi. NW. Yreka	4,260	160	XXI, p. 480
New Era Placer (Lower Siskiyou)					1,000	19	XXVII, p. 60
New Hope Placer (Mattoon)	3	44 N.	9 W.	R. A. Mattoon, Fort Jones	3,200	77	XII, p. 289; XIII, p. 417; XXI, p. 452; XXVII, p. 47. See text this report
New York	2	44 N.	9 W.	Geo. A. Milne, Fort Jones	3,300		text this report
Nirolett		15 N.	7 W.	4 mi. S. Sawyers Bar			XIV, p. 289; XIII, p. 417
Nigger Boy	20, 21, 28, 29	46 N.	8 E.		4,000		XIV, p. 837; XXI, p. 452; XXVII, p. 47. See text this report
Nigger Hill Placer		10 N.		Bennett Co., Forks of Salmon	1,400	80	XII, p. 289; XIII, p. 417; XXI, p. 481
Ninety (Hibernia)				3 mi. S. Sawyers Bar	3,350		XII, p. 284; XIII, p. 417
Nolan Gulch Placer	14	39 N.	9 W.	Charles Crutchfield, Callahan	6,000		XXI, p. 481
Noral Min. Co.		39 N.	11 W.	Near Rollin	5,000	400	See text this report
Nordheimer Placer	3	10 N.	7 E.	Nordheimer Mining Co., H. B. Morey, Sec., Menlo Park	1,100	60	VIII, p. 611; XIV, p. 854; XXI, p. 481
Norma	28	46 N.	8 W.	Geo. J. Cottle, Walker	3,000	240	See text this report
Northern California Dredge Co.	36	46 N.	11 W.		1,500		Bull. 57; XIV, p. 864; XXI, p. 490
Northern Calif. Goldfields				See Morrison & Carlock			
North Star				See Boyle			
Oak Bar Dredge				Near Oak Bar	1,800		XXI, p. 491
Oak Bottom Placer Syndicate	21, 32; 35, 36	12 N., 11 N., 11 N., 12 N.	7 E., 7 E., 6 E., 6 E.	Robert L. Younger, Somes Bar, California, or Medford, Oregon	600	480	XXI, p. 481
O'Connell's Gold Mines, Inc.	1, 2	46 N.	6 W.				See text this report
Ohio				7 mi. W. Yreka	2,500	60	XIV, p. 838; XXI, p. 453
Ohio		9 N.	8 E.	A. Nally	2,900		XIII, p. 418
Old Indian	4				3,000	40	XIV, p. 838
Old Vet and Eclipse (Hossick & Brown)				7 mi. N. Fort Jones	4,000		XIII, p. 418; XIV, p. 838; XXI, p. 453
Olive	10, 11	45 N.	8 W.	5 mi. NE. Sawyers Bar, Sam Wallace, Frank Smith, Wm. Perkins, Sawyers Bar	6,500		XII, p. 289; XIII, p. 419
One Hundred Dollar							XXI, p. 453

Name	Operator / Location	Sec.	Twp.	R.			References
O'Neill Ranch Placer	B. J. Getchell, Hamburg Bar	22, 27	46 N.	11 W.	1,500		XXI, p. 481
Orcutt Placer	Alvin Orcutt, Forks of Salmon	30	39 N.	12 W.	1,500	20	XXI, p. 482
Oregonian	See Malloy						
Oregonian	4 mi. N. Sawyers Bar		45 N.	8 W.	3,800		XII, p. 289; XIII, p. 419
Oregon (Hegler & Aldrich)	Ed. Mathewson	3	45 N.	8 W.	3,900		XI, p. 446; XII, p. 284; XIII, p. 405; XXI, p. 453
Oriental	½ mi. S. Rollin, John Joseph, Sawyers Bar						XXI, p. 453
Oro Grande (Cummings, McKeen or Shasta)	Oro Grande Mining Co., Hugh McKinnie, Pres., Callahan	36	40 N.	9 W.	4,000	480	XII, p. 280; XIII, p. 396; XXI, p. 454; XXVII, p. 48. See text this report
Oro Grande	Trask & Corinson	10, 11	45 N.	8 W.	3,500	100	XIII, p. 419; XIV, p. 838; XXI, p. 454
Osceola (Franks & Moncton)	5 mi. SE. Sawyers Bar, James Keaton, Sawyers Bar						
Osgood (Lucky Strike)	J. E. Hubbart, c/o J. C. Hubbart, Yreka	16, 21	45 N.	7 W.	4,500	60	XII, pp. 282, 289; XIII, p. 419; XXI, p. 454
Outlook	See Sheba				2,800	520	XIII, p. 138; XXI, p. 454; XXVII, p. 44
Overton	Overton Gold Mining Co.	16	40 N.	10 W.	4,000		XIV, p. 838; XXI, p. 455
Paddy Cronin	See Gallia						
Paine Placer	In Scott Bar District, Wm. Fridley, Scott Bar						XXI, p. 482
Paradise Flat Placer	Robert Younger	29	40 N.	11 W.	2,100	40	XIV, p. 863; XXI, p. 482
Paradise	W. F. Davis, 427 J St., Sacramento	18	15 N.	7 E.		460	See text this report
Paragon Placer	E. E. Friend, Happy Camp	32	17 N.	8 E.		20	
Parker	George J. Parker, Copper, Oregon	20 or 21	45 N.	11 W.	2,100	40	XXI, p. 482
Paymaster	4 mi. S. Sawyers Bar				4,300		XIII, p. 419
Pennsylvania	See Fawcett						
Peters	6 mi. S. Sawyers Bar				4,725		XII, p. 289; XIII, p. 420
Peterson Placer	See Joubert						
Phillips & Cartwright	8 mi. W. Yreka						XII, p. 290
Piedmont	Gertrude G. Chick, 1070 Pine St., San Francisco. Mine near Cecilville						
Pilot Knob and Kearsarge (Midwinter & Modoc)	R. G. Reeder, Yreka	30	46 N.	6 W.	3,000	40	XVII, p. 535; XXI, p. 456
Pilot Knob	Pilot Knob Mining Co., T. K. Anderson, Sec., Gottville	23, 24, 26	47 N.	7 W.			XIV, p. 839; XXI, p. 455
Pine Grove	Evans Gulch in Quartz Valley	10	46 N.	9 W.	3,000	310	XIV, p. 854; XXI, p. 482
Pinkham	½ mi. W. Oro Fino				2,000		XIII, p. 420
Pitts	1 mi. S. Callahan						
Poker Flat Dredge	H. H. Albars, Blue Nose	20	13 N.	6 E.	3,400		XII, p. 290; XIII, p. 420
Pond Water Placer	Mr. and Mrs. Chas. Slater, Callahan	18	40 N.	7 W.	600	20	XXI, p. 491
Porphyry Dike	R. V. Hayden, F. D. and R. C. Sullivan, Callahan	6, 7	40 N.	8 W.	4,000	20	XXI, p. 482
Porters Bar Dredging Co.	See Humpback						XXI, p. 456
Portuguese Quartz	Henry Wood, Seiad, see Minerals Recovery Corp., Ltd., also	4	46 N.	12 W.	3,000	350	XXI, p. 491
Portuguese Placer							VIII, p. 595; XII, p. 290; XIII, p. 420; XIV, p. 420; XIV, p. 855; XXI, p. 482
Poverty Point Placer	H. J. Barton, Yreka	18	46 N.	9 W.	1,500	200	VIII, p. 592; XIII, p. 420; XIV, p. 863
Prospect Hill		30	12 N.	6 E.	1,800		XXI, p. 456; XIV, p. 839
Providence		18	43 N.	9 W.			XXI, p. 421; XIV, p. 839; XXI, p. 456
Punch Creek	5 mi. NW. Yreka				3,000		XIII, p. 421
Quartz Gulch Placer	See Wingate Hill				2,850		
Quartz Hill	Harry G. Noonan, 2801 Jackson St., San Francisco	16	45 N.	10 W.	1,900	75	VIII, pp. 622, 623; XI, p. 447; XII, p. 290; XIII, p. 421; XIV, p. 839; XXI, p. 453; XXVII, p. 49. See text this report

TABLE OF GOLD MINES AND PROSPECTS, SISKIYOU COUNTY—Continued

With References to State Mineralogist's Reports

Name of mine	Sec.	Twp.	Range	Remarks on location or owner	Elevation, feet	Area, acres	References
Queen of Sheba							
Queen of Sheba							
Rainbow	10	40 N.	9 W.	10 mi. SW., Yreka.	4,700		XIII, p. 421
				Abel Goldberg, Etna.			
Red Hill	4	45 N.	7 W.	Near Hawkinsville, T. C. Quinn & Newton	4,000	250	See text this report
Red Hill Hydraulic				Gordon, Hawkinsville.			
Reeder							
Reeve Ranch Placer	2, 11	46 N. 16 N.	7 W. 7 E.	At Gila	3,500		XVIII, p. 733; XIX, pp. 58, 93; XXI, p. 456
Reliable and Extension				Miss M. A. Reeve, 720 Oak St., San Francisco	2,600	160	XIV, p. 839; XXI, p. 456
Reno					1,200	281	XIV, p. 615; XXI, p. 456
Renown	14	44 N.	9 W.	On Hungry Creek, 9 mi. from Coles Station			VIII, p. 697; XI, p. 443; XIII, p. 291; XXI, p. 484. See text this report
Richardson	1	41 N.	7 W.	N. J. Rowley, Fort Jones	3,000	40	XII, p. 291; XIII, p. 422
Riverside Placer	16	17 N.	7 E.	James Furlong, Gazelle	5,000	480	See text this report
Roberts	32		8 E.		1,100	450	XXVII, p. 52
Robinson Gulch	18	40 N.	10 W.	E. D. Friend, Happy Camp.	3,600	20	See text this report
Rocky Point Placer	32	45 N.	10 W.	4 mi. S. Sawyers Bar	2,500		XII, p. 291; XIII, p. 422
Roff (Cory or Bridger)	34	41 N.	9 W.	Hicks & McCarthy, Scott Bar	1,800		XIV, p. 840; XXI, p. 456
Ronning	19	40 N.	9 W.	2 mi. NW. Cecilville	2,680		XXI, p. 484
Roff				V. W. (Dan) Roff, Etna			XIII, p. 423
Rough & Frye	30	12 N.	6 E.	Southern Pacific Co., San Francisco.	3,600	80	See text this report
Rough and Ready Placer				C. Frye			XIV, p. 457
Rout				Harry Walker, Clear Creek P.O.	3,600	40	XIV, p. 840
Roxbury Placer				3 mi. NW. Scott Bar, Roxbury Syndicate, c o Geo. Milne, Fort Jones.	1,600		XXI, p. 423
Ruby Basin				See Jumbo			XXI, p. 484
Rulon Placer	23, 24	40 N.	12 W.	On Beaver Creek, Percy Rulon	2,000	80	XXI, p. 485
Russell Placer	14	39 N.	9 W.	Geo. Russell, Blue Nose P.O.	6,000	60	XII, p. 276; XIII, p. 386; XIV, p. 835; XXI, p. 485
Russian Hill Placer (Red Hill)	14	43 N.	10 W.	Estate of A. Ahlgren, Sawyers Bar	3,500	20	XII, p. 423; XIV, p. 835; XXI, p. 485
Rycroft Placer	12	39 N.	10 W.	S. Rycroft.	6,600		See text this report
Saint Lawrence		39 N.	8 W.	A. G. Myres, Mines Co., Callahan			XXI, p. 485
Salmon River	15			Salmon River Mines & Development Company	1,400		XXI, p. 291
Salsbury & Edmundson Placer				See Siskiyou Mines & Development Company	1,100	40	XIV, p. 835; XXI, p. 485
Sanford & Edmundson Placer				Near Seiad, J. A. Sanford & T. A. Edmundson, Seiad			
Santa Teresa	34	11 N.	7 E.	On Hungry Creek	5,500		VIII, p. 626; X, p. 656; XI, p. 446; XII, p. 291; XIII, pp. 424, 425, XXI, p. 437; XXVII, p. 51. See text this report
Sauerkraut Placer	17	45 N.	8 W.	Charles Lilly, Forks of Salmon	1,700	20	
Schroeder			10 W.	Fidelity Metals Corp., c/o R. A. Peabody, 2989 21st Ave., San Francisco.			XIII, p. 425; XIV, p. 965
Schuler Drift Placer	16	45 N.	10 W.	M. Schuler			

Name	Sec.	T. N.	R.	Location and remarks			References
Schuler Placer	12	16 N.	7 E.	2 mi. NE. Happy Camp	1,100	60	XXI, p. 485
Scroc	13	43 N.	10 W.	S. F. Bentley et al., Fort Jones	3,500		XXI, p. 457
Scott Bar				See Quartz Hill			
Scott River Dredging Co.	11	46 N.	12 W.	Near Callahan	1,400	60	Bull. 57, p. 221; XIV, p. 864
Seiad Placer				T. K. Towne			XIV, p. 856
Seiad Placer Mines				See Lowden (Ariel Lowden)			
Shadows Creek Placer	4	15 N.	7 E.	Near Cecilville	2,000	80	XXI, p. 485
Sheba (Outlook)				Peter Grant & Alfred Effman, Happy Camp			XXI, p. 455; see text this report
Sheffield				4 mi. E. Sawyers Bar			XII, p. 291
Short Bend Placer	18	16 N.	8 E.	Forest Moore & C. G. Lewis, Happy Camp	1,100	40	XII, p. 292; XIII, p. 426
Short				10 mi. E. Cecilville			See text this report
Shumway Placer (Bonaly)				1 mi. E. Forks of Salmon River Mining Co., Flatiron Bldg., San Francisco.			XII, p. 292; XIII, p. 426; XXI, p. 485
Siskiyou County Placer	31	18 N.	7 E.	Owner: Siskiyou County	1,200	220	See text this report
Siskiyou Dredging Co.	6, 25, 14	44 N., 44 N.	9 W., 8 W.		1,900	100	See text this report
Siskiyou Gold Mines Co.				On Horse Creek 8 mi. above mouth. Address: 2045 E. 70th St., Chicago, Illinois	3,000	165	Bull. 57, p. 223; XIV, pp. 864, 865; XXI, p. 491
Siskiyou Klondike Placer				In Oak Bar District	2,500	80	XXI, p. 457
Siskiyou Metals Co.						40	XIV, p. 856
Siskiyou Mine				7 mi. NW. Yreka	2,950		See text this report
Siskiyou Mines & Development Co. (Placer)	5, 6, 7, 18	39 N.	8 W.	A. C. Aiken, San Francisco	1,200	1,500	XIII, p. 426
Siskiyou River Bend Placer				18 mi. W Callahan, A. E. Westover, Callahan, or 480 Pine St., San Francisco	1,700	20	XIV, p. 856; XXI, p. 486
Six-Mile Creek Placer				D. Skelton			XXI, p. 487
Skelton	26	41 N.	9 W.	5 mi. E. Sawyers Bar, G. T. Salsbury, Berkeley	3,700	40	XXI, p. 487
Slim Jim				Bennett Co., Forks of Salmon	4,000	20	XIV, p. 840
Slumway Placer	3	10 N.	7 E.	4½ mi. SE. Sawyers Bar	3,000	20	XXI, p. 457 See text this report under "Lanky Bob"
Smith & Merriam				On French Creek	1,150	40	XIV, p. 856
Smith					3,360		XII, p. 292; XIII, p. 427
Snowflake (Bear Den) and Little Gem	29	39 N.	3 W.	4½ mi. SE. Sawyers Bar, James Cavanaugh, Rollin	3,000		XII, p. 292
Soda Mint				J. J. Murphy, Weed	3,500	40	XXI, p. 457
Spears	15	40 N.	9 W.	See Norcal Min. Co.			See text this report
Specimen	1	40 N.	9 W.	W. Ellis	3,500	60	XIII, p. 427; XIV, p. 840
Spencer (Goodenough)				12 mi. NW. Yreka	4,500		XI, p. 445; XIII, p. 427
Squaw Gulch Placer	12	43 N.	10 W.	D. L. Jones	3,420		XIII, p. 427; XIV, p. 857; XXI, p. 488
S. R. Crary (Quartz and Placer)				S. R. Crary, Walker P.O.	3,500	80	XXI, p. 487
Star				F. Star	3,550	20	XII, p. 428; XIV, p. 841
Steel				Evans Gulch, Quartz Valley			XII, p. 292; XIII, p. 428
Stenshaw Placer	28, 29	13 N.	6 E.	Samuel Stenshaw, Blue Nose P.O.		98	VIII, p. 604; XXI, p. 487
Sterling Gold Mining Co.				See Golden Eagle & Schroeder			XI, p. 446; XIII, p. 428
Sterling				9 mi. W. Cole's Station			XIII, p. 428; XIV, p. 841
Sterling				13 mi. NW. Fort Jones			XIV, p. 857
Sturn (Colby)	2	16 N.	7 E.	G. H. Compton	1,200	200	
Sugar Creek Mining & Milling Co.				See Hathaway			
Sugar Hill (Quartz and Placer)	36	40 N.	9½ W.	5½ mi. SW. Callahan.	4,800	40	VIII, p. 611; XIV, p. 864; XXI, p. 488

TABLE OF GOLD MINES AND PROSPECTS, SISKIYOU COUNTY Continued

With References to State Mineralogist's Reports

Name of mine	Sec.	Twp.	Range	Remarks on location or owner	Eleva-tion, feet	Area, acres	References
Sundown	19	47 N.	7 W.	Denver Mining Co., C. A. Von, Pres., Denver		60	XIV, p. 841
Sunnyside Placer	1	16 N.	7 E.	H. J. Barton, Yreka	1,100		See text this report
Sunshine				Sucker Gulch, Quartz Valley	3,110		XIII, p. 429
Superior Consolidated Mines Co.				See Johnson & Lewis, Morrison & Carlock, and New York			
Swedish American Mining Co.	30	11 N.	8 E.	See Commodore		40	XIV, p. 841
Taft	18	47 N.	7 W.	C. Taylor	2,900	40	XIV, p. 841
Teddy-Avalon	28	12 N.	6 E.	E. J. Durkee		80	VIII, p. 605; XIII, p. 430; XIV, p. 857; XXI, p. 488
Ten Eyck Placer Mines, Inc.	5	39 N.	11 W.	W. T. Tyler, Sec., 715 Bryant St., Palo Alto	2,600	20	XIV, p. 857
Thomain Placer	29	14 N.	6 E.	F. and C. F. Thomain			XXI, p. 488
Thomas Placer	15	18 N.	7 E.	Thomas Bros., Blue Nose P.O.			See text this report
Thompson Creek				S. K. Wood, Seiad	3,250		XI, p. 437; XIII, p. 430
Tiger	12	39 N.	10 W.	Hull Gulch, Quartz Valley			
Trail Creek				Trail Creek Mining Co., Geo. A. Foster, Pres., Callahan	7,200	100	XIV, p. 841
Trust Buster	10	47 N.	8 W.	J. L. Corbett, Hilt	4,500	160	See text this report
Turk	12	43 N.	10 W.	Banks & Maginnis, Inc.	3,500	40	XIV, p. 842
Twan & Hannan	30	12 N.	6 E.	Twan & Hannan			XIV, p. 857; XXI, p. 488
Two and a Half (Walker Placer)	20	40 N.	7 W.		4,000		
Uncle Sam Consolidated Quartz Mine	3, 10	39 N.	11 W.?	S. P. Tillman	5,000		VIII, p. 619; XI, p. 433; XII, p. 263; XIII, p. 431; XIV, p. 842; XXI, p. 458
Uncle Sam				12 mi. NW. Yreka	3,600		XIII, p. 430
Underland	28	39 N.	3 W.	J. J. Murphy et al., Weed	4,000	120	See text this report
Union (Central or Evening Star)				1 mi. S. Rollin, Rollin Mining Co., c/o Ben Daggett or Leslie Daggett, Black Bear	3,900	40	VIII, p. 620; XII, p. 251; XIII, p. 400; XXI, p. 458. See "Norcal" also
Vesa Creek				12 mi. NW. Yreka, Harry, Joseph and E. Davidson			XXI, p. 459
Victoria	6	39 N.	3 W.	L. H. Brown, Dunsmuir	4,000		XXI, p. 535; XXI, p. 459
Victory Gold Mines	16, 17, 20	40 N.	10 W.	John Nefroney, Etna	4,000	220	XXI, p. 460. See text this report
Volcano				Evans Gulch, Quartz Valley	3,200		XII, p. 293; XIII, p. 431
Wanda				7 mi. SE. Forks of Salmon, O. L. Palmer, Forks of Salmon	2,500	60	XXI, p. 461
Welch & Bradley Placer				On Beaver Creek			XXI, p. 489
Wescoatt				See Norral Min. Co.			
Whistle Bar Placer	29	40 N.	11 W.	Ed. Kearns, Sawyers Bar and Robert Kearns, San Jose	2,100	20	VIII, p. 616; XIV, p. 864; XXI, p. 489
White Bear					4,000		XIV, p. 842. See text this report
White Dog							See text this report
White Elephant		39 N.	12 W.	1 mi. S. Hansen Mine	4,140		XIII, p. 431

Name	Sec.	Twp.	Rge.	Owner or location	Elev.		References
Whittaker	2	16 N.	7 E.	See Blue Gravel		80	See text, Blue Gravel
Wicks				12 mi. NW. Fort Jones	6,200	40	XIV, p. 842
William Burns Placer				See Joubert			XXVII, p. 54
Williams	32	40 N.	9 W.	R. D. Williams, Callahan	6,000		
Williams Point Placer	17	16 N.	8 E.	Mrs. M. A. Fowler, Happy Camp			
Willard Placer				See Hoosier Hill			
Wilson Bros				3½ mi. S. Sawyers Bar, Wilson Bros, Sawyers Bar			
Wilson Placer	29	45 N.	8 W.	F. L. Wilson	3,500	40	XXI, p. 461
Windeler Placer	20	40 N.	11 W.	J. C. Windeler	3,500		XIII, p. 432; XIV, p. 864
Wingate Hill Placer	5, 6	15 N.	7 E.	Anderson & Van Hoy, c/o O. Y. Anderson, Happy Camp	3,000	20	XIV, p. 858
					1,000		VIII, p. 601; XII, p. 284; XIV, p. 855; XXI, p. 483. See text this report
Winterings				See Big Cliff			
Wood & Fehely	4	46 N.	12 W.	Wood & Fehely	1,300	60	XIII p. 432; XIV, p. 858
Woodfill & Barry Placer				1 mi. S. Sawyers Bar, J. Woodfill and C. J. Barry, Sawyers Bar	2,150	80	XXI, p. 489
Wood				On Thompson Creek, 10 mi. above mouth. Sam K., Stanley and Mrs. J. C. Wood, Mrs. Lena Hibbard and E. L. Brier, Seiad			
Wright & Fletcher	7, 8	43 N.	9 W.	Wright Bros.	3,000	120	XXI, p. 461
						80	VIII, p. 609; XII, p. 284; XIII, p. 433; XIV, p. 858
Yellow Jacket	20	37 N.	9 W.	7 mi. W. Walker.			XIII, p. 433
Yellow Rose				McCormick Saeltzer, D. V. Saeltzer et al., Redding	6,700	60	XXII, p. 26; XXVII, p. 55. See text this report
Zarina	33	41 N.	10 W.	Zarina Mining Co., J. W. Harris, Supt., Etna	6,500	100	XIV, p. 842

GRANITE.

Good quality granite occurs on Craggy Mountain in T. 41 N., R. 8 W., and in Sec. 1, T. 41 N., R. 9 W., also in T. 40 N., Rs. 8 and 9 W. This has been used locally in a small way for building blocks and monuments. The deposits are all a long distance from railroad.

Bibl: Cal. State Min. Bur. Bull. 38, p. 54.

IRON.

Float iron ore is mentioned as occurring in Sec. 10, T. 46 N., R. 10 W.

Bibl: Cal. State Min. Bur. Bull. 38 p. 304.

LEAD.

On the *Fippen & Hayden Placer* (see under Gold Placer Mines), in the Callahan district, an adit of unknown length was run, many years ago, to prospect for lead. A small amount of good galena ore was brought out of this adit and dumped. In 1924, M. E. Gardner sluiced off the soil overburden and exposed the edge of the vein in place above the old adit. The exposed section is about 30 feet long and the width of vein is from a few inches to one foot. It consists of barite, carrying lumps and stringers of galena. Assays of the ore indicate a high lead content, sufficient to make the best samples a good shipping ore, and with a fair content of silver. Ore could be concentrated by jigging. Gazelle, the nearest railroad point, is 27 miles distant by road. This prospect is idle.

Other prospects of lead ore have been reported in the region eight to ten miles east of Callahan near the Gazelle road, but so far as could be learned, the float found there had not yet been traced to the vein, although some prospecting was going on in the late summer of 1925.

Siskiyou Lead Mine, the patented NE¼ Sec. 29, T. 41 N., R. 7 W., 15 miles west of Gazelle, is the property of M. H. Balfrey of Gazelle. Evans and Mitchell were leasing at the time of visit. A crosscut adit cuts the vein at a distance of 70 ft. from the portal, giving a depth of 35 feet. Galena shows in spots in a vertical mineralized zone reaching a width as great as 12 ft. in places. A drift has been driven on this for 100 ft. Two shafts on the vein, reaching to depths of 50 ft. and 80 ft. below the surface were full of water to the tunnel level. Lessees state that 600 tons of ore on the dump assay 8% lead, 2% zinc, and $1.50 per ton in gold. The country rock is andesitic, some of it coarse enough in grain to approach diorite. A small mill manufactured by the Sacramento Gold Mill Co. was being installed. It consists of four stamps, of about 35 lbs. each, mounted on springs to give force to the blow. A jaw crusher, 6 in. by 8 in., will crush the feed for the mill. There are also two concentrators, 5 ft. by 7 ft., similar to Wilfley tables, but made by the same company as the mill. A 30-hp. gasoline engine will drive the outfit.

LIMESTONE

Large deposits of limestone and marble occur in Siskiyou County, but many of them are remote from present railroads. Those nearest to transportation are in the area of Devonian rocks just west of Gazelle. For local use in Scott Valley, deposits near Etna Mills, Greenview and Callahan have been worked in a small way. An area of Paleozoic and possibly pre-Paleozoic rocks with outcrops of limestone here and there extends from near Callahan northward for some 25 or 30 miles.

Marble occurs in large deposits in a belt running west of Etna Mills along the east slope of the Salmon Mountains, the Grider Mountains and past Thompson Creek. Should the demand ever arise, and means of transportation be provided, Siskiyou County could supply nearly any demand for limestone. The two deposits described below are the only ones from which production has recently been reported.

Hathaway Limestone Quarry, owned by Hubert Hathaway of Etna Mills, is in Sec. 9, T. 41 N., R. 8 W. The limestone outcrops on patented agricultural land in an area of about two acres to a height of 60 to 100 feet, and was mined in an open quarry. The last limestone, about 300 tons, was ground in 1929, under the supervision of A. S. Hathaway, who has since died. Dry grinding was done in a single stage to a maximum size of ⅛ inch in a Day hammer-mill, made in Knoxville, Tennessee. A Fordson tractor furnished the power. Ten to 15 tons were ground per 8-hour day with a crew of four men, including a truck driver. Prices received were $4 per ton at the quarry, or $6 per ton delivered to points in Scott Valley. Hubert Hathaway does not plan any further production, and the hammer-mill is for sale. This is the same quarry from which Wm. Werst of Etna Mills produced limestone a few years ago. No analysis of this limestone is available. It is probably very similar to that of the Mt. Shasta Lime Co., an analysis of which is given below.

Mt. Shasta Lime Company is a partnership of W. J. Chastain of Gazelle and H. A. Craig. The property, four miles west of Gazelle by a good dirt road, comprises 760 acres in Sec. 7, T. 42 N., R. 6 W., and Sec. 12, T. 42 N., R. 7 W. Of this 460 acres are patented and the balance held by mineral locations. The limestone is mined in an open quarry and burned in a vertical kiln burning wood. The capacity is seven tons per day of 24 hours, and seven men are required when the kiln runs steadily. An average of 200 to 300 tons of lime per year is sold, but the kiln has been operated as much as seven months in one year. The last lime sold brought $20 per ton f.o.b. Gazelle packed in boxes holding half a barrel each. Cost of boxes was 44¢ per bbl. In the adjoining Sec. 6, T. 42 N., R. 6 W., W. J. Chastain and B. B. McCoy own 270 acres of land on which there is another large limestone deposit.

The following analysis from the above quarry was furnished by W. J. Chastain, who stated that it was made by Smith-Emery, No. 69,069:

Silica (SiO₂)	0.18%
Alumina (Al₂O₃)	0.10
Iron Oxide (Fe₂O₃)	0.20
Lime (CaO)	55.26
Magnesia (MGO)	0.27
Loss on ignition	43.76
	99.77%
Calculated (CaCO₃)	98.65%

Following are the locations and ownerships of some of the larger limestone deposits of Siskiyou County which are fairly close to existing roads:

T. 41 N., R. 8 W.

Sec. 9. See under Hathaway above.

Sec. 20, SE¼, Public domain.

Sec. 29, N½, Southern Pacific Land Co., Southern Pacific Bldg., San Francisco.

Sec. 32, NE¼, Public domain.

Sec. 33, NW¼, Southern Pacific Land Co., Southern Pacific Bldg., San Francisco.

T. 42 N., R. 6 W.

Sec. 6, SE¼, Geo. A. Stockfleth, c/o Sullivan, Roche, Johnson & Berry, 220 Bush Street, San Francisco.

Sec. 6, Balance, public domain.

Sec. 7, NW¼, H. A. Craig, Gazelle.

Sec. 7, SW¼, Southern Pacific Land Co., Southern Pacific Bldg., San Francisco.

Sec. 18, NW¼, Robert M. Hamilton, Gazelle.

T. 42 N., R. 7 W.

Sec. 1, SE¼, Clarance McCoy, Gazelle.
NE¼, Southern Pacific Land Co., Southern Pacific Bldg., San Francisco.

Sec. 7, See under Mt. Shasta Lime Co. above.

Sec. 12, NW¼, Barney McCoy, Gazelle.

Sec. 12, See also under Mt. Shasta Lime Co. above.

T. 43 N., R. 8 W.

Sec. 35, SE¼, Southern Pacific Land Co., Southern Pacific Bldg., San Francisco.

Sec. 36, W½, Frank Sharp, Fort Jones.

T. 44 N., R. 7 W.

Sec. 4, N½, Clara Walter and William Murray, Yreka.

Deposits that have been listed in past reports follow:

Name of deposit and location   *Remarks*

Burton, Sec. 17, T. 43 N., R. 9 W., 4 mi. NE. of Greenview____limestone

Farrington, Sec. 24, T. 40 N., R. 9 W., 2½ mi. SW. of Callahan_limestone

Marble Mountain, Sec. 16, 19, 20, 21, T. 43 N., R. 10 W._____marble

Barton & Barham, Sec. 18, T. 46 N., R. 9 W., in Oak Bar district.

Luce, two miles north of Fort Jones in Deadwood district__good marble

McDaniels, Sec. 6, T. 41 N., R. 9 W._____white marble

Parker, Sec. 32, T. 42 N., R. 9 W., near Etna Mills___pink, white marble

Thompson Creek, Sec. 8, T. 17 N., R. 8 E., H. M., 8½ mi. NE.
of Happy Camp _____white marble

Bibl: (Siskiyou County limestone) State Mineralogist's Reports XIII, p. 632; XIV, pp. 865, 867; XXI, p. 492; XVII, map Cal. State Mining Bur. Bull. 38.

MANGANESE

Manganese prospects have been reported from Sec. 9 or 16, T. 44 N., R. 8 W., southwest of Yreka; Sec. 15, T. 46 N., R. 6 W., six miles southwest of Klamathon; Secs. 20 and 21, T. 43 N., R. 9 W., on the ridge north of Greenview and southeast of Oro Fino; and in T. 39 N., R. 11 W., near the head of Callahan Gulch. For further particulars the reader is referred to California Division of Mines Bulletin 76, 'Manganese and Chromium,' published in 1918.

A manganese prospect was reported, about 1931, as being found in grading the new road from Black Bear to the King Solomon mine.

MINERAL WATER

There are in the county a number of mineral springs, several of which have been known for years as summer resorts. Considerable water has also been shipped from Shasta Springs. Among the principal resorts are Shasta Springs, Warmcastle Soda Springs, Klamath Hot Springs, Garretson Springs, Neys Springs and Upper Soda Spring. These, except Klamath Hot Springs, are cold carbonated springs. All of these springs have been described in the past publications of the Bureau and at a later date (1915) in U. S. Geological Survey Water Supply Paper 338, which covers the subject in great detail, so that there is no need of repeating this information here. Waring in the last mentioned report lists 15 groups of carbonated springs in the county. They may be grouped in three localities; those near Beaver Creek, in the northeastern part of the county, those in the western part of the lava-covered area in the vicinity of Ager, and a third and most important group because of their accessibility, near Sacramento River in the southern part of the county.

Bibl: Cal. State Min. Bur. R. XIII, p. 520; XI pp. 449, 452. U. S. G. S.; W. S. P. 338; Winslow Anderson, Mineral Springs of California.

MOLYBDENITE

Yellow Butte Mine. See under Copper.

ORNAMENTAL AND GEM STONES

Californite, a massive green vesuvianite, was first found on Indian Creek, north of Happy Camp. It is a hard, massive mineral with a vitreous luster and varies in color from nearly white to olive green. It has been mistaken for jade here and in other localities. The mineral occurs on claims belonging to the *Reeves Davis Estate* and *Henry Howard*, on the South Fork of Indian Creek, 10½ miles north of Happy Camp. It also is found as float, in large and small boulders, on placer claims belonging to *Scott Greening*, on O'Meara Creek, a tributary of Indian Creek, which enters it at the Classic Hill Mine, 12 miles north of Happy Camp, and pieces of float have been found elsewhere along the course of Indian Creek and in the mountains on the west side of the creek. Nothing has been done to develop it.

Bibl: Cal. State Min. Bur. Bull. 37, pp. 93, 94; R. XIV, p. 869.

Onyx occurs 200 yards downstream from Shasta Springs near Sacramento River. The deposit has been known as the *Griffin Onyx Marble Quarry* and contains a vein about five inches wide. Small articles such as paper weights have been made from it. Idle.

Fire opal is mentioned by Eakle * as occurring near Dunsmuir.

C. B. Kay of Ager has lately been doing some work on a deposit of fossiliferous sandstone which occurs on the *Henry Hagedorn* and *Sidney Richardson Ranches* five miles south of Ager in the NE¼ of Sec. 26, T. 46 N., R. 6 W. There are several beds of the stone, ranging from three inches to one foot wide and dipping east about 23°. Some of these beds are so thickly matted with shell remains as to give the polished stone a very pleasing and unique appearance as the sandstone is nearly black and in contrast with the white shells. The stone being fine-grained is hard and compact enough to take a polish, and might find a market for small articles. It contains a high percentage of cementing calcite in addition to that in the fossil shells.

PLATINUM GROUP METALS

Osmiridium and platinum occur with gold in the gravel deposits along Klamath River and Salmon River. The subject of the occurrence of these metals was covered by C. A. Logan in considerable detail in the Bureau's Bulletin 85. These metals occur in very small proportion to the gold content of the gravel, and the output from the county is negligible.

Field work for the above report showed that while platinum group metals occur in the gravel of the South Fork of Salmon River in appreciable amount from Matthews Creek downstream, at least as far as Forks of Salmon, there has been practically no platinum or osmiridium found in the placer mines of the North Fork, although these have yielded several times as much in gold as the placers of the South Fork. The presence of platinum group metals on the South Fork side, and their absence from the other fork, is attributed by the writer to the presence of considerable outcropping serpentine and peridotite in the drainage basin of the South Fork, while these rocks are found only in a few very small areas on the North Fork side. The platinum and osmiridium formerly recovered at the Nigger Hill and Michigan-Salmon hydraulic mines probably came from nearby serpentine areas. Nuggets up to ½ ounce in weight were found. An analysis of a shipment from the latter mine showed 97.05% osmiridium and 2.95% platinum. This comes nearer to being pure osmiridium than any other known sample in the state. While platinum usually occurs in the placers as thin, smoothly hammered flakes, similar in size and shape to fine placer gold, on account of its softness, osmiridium or iridosmine usually are found in ragged, angular pieces because of the great hardness of these native alloys. The latter may be dull, rusty, or coated, and might easily be mistaken for lead were it not for the fact that none of the metals of this group can be melted in an open fire as can lead, and the above named alloys are much harder than lead and as a rule so feebly magnetic (if at all so) that they are easily distinguished from iron or its compounds.

---

* Eakle, A. S., Minerals of California; Cal. State Min. Bur. Bull. 91, 1923.

One ounce of platinum group metals to $3,000 in gold is about the usual proportion in this state where the two occur together. The proportion of platinum is a little higher than this on the South Fork of Salmon, and lower on the Klamath. None of the other placers of the county are known to have produced platinum, and none was recovered from the dredging operations on McAdams and Greenhorn creeks.

It is believed that the platinum metals in place occur only in narrow stringers or seams in serpentine and in the rocks from which serpentine is derived, such as peridotite, dunite and gabbro. Only one report has come to the writer's attention of platinum in place in this county. This was in a quartz stringer prospected by Morgan Bros. of Gottville, while hunting pockets on Empire Creek. They state that the report was made by a Denver assayer. The quartz prospect gave out entirely after being followed a few feet.

PUMICE

G. Z. Johnson, 255 California Street, San Francisco, produces pumice from Pumice Stone Mountain in T. 43 N., R 2 E., M. D. M., near Little Glass Mountain southwest of Mt. Hoffman.

Mt. Hoffman Pumice Claims comprise 960 acres, being all of Sec. 28 and the E½ Sec. 29, T. 44 N., R. 4 E. at Mt. Hoffman, in eastern Siskiyou County. These placer claims are held by E. L. Jameson, c/o S. C. Pennell, Tennant, California, Dan A. Williams, 418 California St., Salinas, California, and J. O. Miller. The property is 35 miles by road east of Tennant. A road has been built for this entire distance, but at time of visit (October, 1935) the last three miles could not be used because parts of the road had been destroyed by extensive logging operations.

Small pieces of pumice of the size of a marble and smaller are abundant in the soil for many square miles in this vicinity. Apparently much of this deposit is superficial, being only a foot or two deep. On the Mt. Hoffman claims, which are high on the mountain at an elevation of 7000 ft., the pumice is free from soil in many places, the lumps are larger, and the depth is greater. Here the larger lumps are several feet in diameter, and it is possible to saw bricks from them. The best pumice occurs on a little ridge, roughly half a mile long, 200 ft. high, and several hundred feet wide. One end of this ridge is capped by a flow of obsidian; but the larger part of it is free of this capping, and the surface is composed entirely of pumice. Tunnels have been driven into the pumice for 30 ft. exposing a mixture of lumps of various sizes, most of them small, but some of them 3 ft. to 4 ft. in diameter. One tunnel is said to have been driven for a distance of 125 ft. in pumice, but no tunnel of this length was open to inspection at time of visit.

A few tons of pumice have recently been shipped by truck for sawing into scouring-bricks, and for crushing to small sizes for use in acoustic plaster. A standard-gage logging railroad of the Long-Bel. Lumber Sales Corporation runs to a point about three quarters of a mile from the pumice deposit.

QUICKSILVER

Cowgill Quicksilver Mine. Owner, Mrs. A. M. Cowgill, P. O. Box 96, Yreka. In Sec. 34, T. 48 N., R. 9 W., on the headwaters of the West Fork of Beaver Creek, 10 miles in an air line from Gottville but farther by road.

According to Bradley * there was some work done here years ago by Siskiyou Quicksilver Mining Company, who built a 10-ton furnace and are said to have produced a few flasks of mercury. During the wartime boom in quicksilver, Mercury Company of America did some prospecting on the surface, and put up a 12-pipe retort. Cinnabar was found in the decomposed surface soil to a depth of 60 feet. This company did little work.

Horse Creek Cinnabar Co., Ltd. The following is quoted from State Mineralogist's Report XXVII, chapter for January, 1931:

"This company is purchasing from H. J. Barton of Yreka 35 acres of patented mining claims in Sec. 15, 16, T. 46 N., R. 10 W., on Klamath River road where the Horse Creek bridge crosses the Klamath. The president is Dr. A. H. Taylor, 5558 Hollywood Boulevard, Los Angeles, and the secretary is Ben Cohn, Horse Creek, Siskiyou Co., California. Mica schist alternates with bands of hornblende schist, the schistosity on the latter not being so well developed as on the former. Small amounts of cinnabar are seen on fractures in the dark-green hornblende schist; and it seems to be associated with this rock only. A mass of light-colored dioritic rock on the property also shows some schistosity. Development consists of three tunnels, 20 ft. to 30 ft. long and numerous (25 to 30) small prospect cuts on the hillside. These average 5 ft. long, 1 ft. wide and 4 ft. deep. The property was equipped with a continuous process retort with a rated capacity of 25 tons per day, which has not yet been successfully operated, because some of the metal parts failed to withstand the heat when it was first fired up. A 25-hp. Fairbanks-Morse gasoline engine furnished the power."

This property has been idle for some time.

Great Northern Quicksilver Mine, Inc. (present name from records of County Assessor) was described in State Mineralogist's XXVII, chapter for January, 1931, as follows:

"*Mercury Mine* (No. 39 on map), formerly the *J. N. P. Mining Co., or Morgan Brothers Mine,* is the property of Eugene Aureguy, Room 333 Mills Bldg., San Francisco. The location is Sec. 13, 14, 24, T. 47 N., R. 8 W., near Gottville or 21 miles from Hornbrook. The first 16 miles are improved highway along Klamath River; and 5 miles are unimproved mountain road. A total of 250 acres are held, also a water right in Sec. 18, T. 47 N., R. 7 W. A fissure vein striking S. 70° W., and dipping 55° to the northwest carries cinnabar and native mercury in a gangue of quartz and calcite. Country rock is schist formed from metamorphosed sediments intruded by granodiorite; and the ore seems to be associated with the contact between these two rocks. However, faulting is in evidence; and the possibility exists that the mineralization has come up along this from younger intrusive rocks below. Less than 10 miles to the east, Tertiary volcanic rocks are exposed on the surface. A stope, 100 ft. long, was being made on the vein mentioned above, with a width of 5½ ft. average, 12 ft. maximum and a height of 20 ft. above the main level. A raise of 140 feet on the vein connects the top of this stope with the surface. This main level is an adit, from which several hundred feet of work, drifts and crosscuts, have been done. A second adit, from which several hundred feet of work have been done, is about 50 feet above and 500 ft. to the east of the main adit. Some ore is exposed here also.

The property is equipped with a flotation plant of a capacity of 24 tons per 24 hours. A crusher, 10 by 12 in., reduces the ore to one-inch size; and this is fed to a 15-ton (rated capacity) ball mill by an automatic feeder, and is here reduced to 40-mesh. This product is treated in three Kraut flotation machines; and concentrates are retorted in Hendy pipe retorts, six pipes, 12 inches by 6 feet, with oil as fuel. An attempt is made to recover native mercury at the discharge of the ball mill on copper plates. A 75-hp. Holt gasoline engine runs the plant. Mine equipment includes a compressor of one-drill capacity, rails and cars. W. W. Young of Gottville is superintendent; and six men are employed.

Bibl: State Mineralogist's Report XXI, p. 496; Bull. 78, p. 169."

In 1935, C. J. Moore, M. J. Orth and F. A. Riggs of Klamath River (postoffice), were leasing the property and were experimenting with a

---

* Bradley, W. W., Quicksilver Resources of California : Cal. State Min. Bur. Bull. 78, p. 169, 1918.

furnace of their own design. It has a circular rotating hearth, 4 ft. in diameter; and the capacity is stated to be 10 tons in 24 hours. At time of visit changes had just been made in the condensing system, and the furnace had not yet been operated with the new condenser of tile pipe, three feet in diameter. These lessees had done no new work in the mine.

SANDSTONE

Sandstone beds form the rim of Shasta Valley, outcrop east of Yreka and are found at several places from Snowdon to the Oregon line (see under Geology). A number of years ago, before cement came into such general use, sandstone was quarried here as well as in other counties of the state. Three quarries were described in our past reports.

Antone or *Weeks Quarry* is two miles northeast of Yreka. The sandstone was quarried in layers from six inches to eight feet thick and has been used considerably for building in Yreka.

Fiock Bros. Quarry is in Sec. 13, T. 45 N., R. 7 W., near Yreka. The stone is coarse grained, even textured and tawny in color.

Southern Pacific Company has quarried considerable sandstone in Sec. 29, T. 47 N., R. 6 W., near Hornbrook. This stone has been used for railroad culverts and for some buildings in Hornbrook.

SOAPSTONE

Soapstone and talc might reasonably be expected to occur at many places in the county, where serpentine is found. For the location of serpentine outcrops, the reader should refer to the sections on Asbestos and Chromite, herein. To date, no commercial development of soapstone has taken place in Siskiyou County. For the most part, the deposits noted are a considerable distance from the railroad. Such deposits, of good size and said to be of satisfactory quality, are found in the northeast part of T. 45 N., R. 11 W., southeast of Hamburg Bar; also near the head of Wooley Creek, and better grades are said to have been found in a deposit on top of the divide between Beaver and Bumblebee Creeks and in a small deposit in Sec. 32, T. 42 N., R. 9 W., near Etna Mills. The business of soapstone and talc production and marketing is highly competitive, and the market prices of lower grade and 'off color' soapstone and talc are so low that only deposits near the railroad can be worked at a profit. There is a scarcity of high-grade, white-grinding talc, however, in the northern end of the state, such ore at present coming from the southern counties, while 'off color' soapstone, used for roofing where the shade is not important, is produced near the railroad in El Dorado County.

STONE INDUSTRY

Since 1916, Siskiyou County has produced considerable crushed rock and gravel for road construction. The stream gravels near Yreka, as for example the dredged portions of Greenhorn Creek, and some areas of volcanic rock near enough the railroad are the most convenient sources of supply. There has been a considerable length of both state highway and rock-surfaced county roads built in Siskiyou

County during recent years, and local stone has been utilized. Part of it was crushed dredger tailing from a point a mile south of Yreka. Buildings of lavas and tuffs may be seen at Mt. Shasta City and Weed. The rocks are crystalline; and the different natural colors and shades, when used together, give very pleasing effects.

SACRAMENTO FIELD DISTRICT

C. A. LOGAN, Mining Engineer

Mr. C. A. Logan, district engineer, is engaged in investigation for a new report upon the gold mines of Placer County.

SAN FRANCISCO FIELD DISTRICT

C. McK. LAIZURE, Mining Engineer

Reports covering the mines and mineral resources of all of the counties in the San Francisco field division are now available, and field work at present is confined to investigations for special reports upon various economic minerals.

LOS ANGELES FIELD DISTRICT

W. B. TUCKER and R. J. SAMPSON, Mining Engineers

Reports covering the mines and mineral resources of most of the counties in the Los Angeles field division are now available, and field work at present is confined to investigations for special reports upon various economic minerals.